ric of
the Cook Islands

ARTISTIC TRADITIONS IN WORLD CULTURES
Series editor: Jonathan C. H. King

Published with the assistance of The Getty Foundation

Body Ornaments of Malaita, Solomon Islands
Ben Burt, with contributions by David Akin
and support from Michael Kwa'ioloa
ISBN 978-0-7141-2578-7

Nomadic Felts
Stephanie Bunn
ISBN 978-0-7141-2557-2

Tivaivai: The Social Fabric of the Cook Islands
Susanne Küchler and Andrea Eimke
ISBN 978-0-7141-2580-0

This new series looks at art and material culture in rapidly
changing societies around the world. Introducing new theoretical
perspectives, each title focuses on how objects contribute to
social relationships through visual and symbolic values specific to
individual peoples.

Susanne Küchler and Andrea Eimke

Tivaivai

The Social Fabric of the Cook Islands

With photographs by Andrea Eimke

THE BRITISH MUSEUM PRESS

© 2009 The Trustees of the British Museum

Published in 2009 by The British Museum Press
A division of The British Museum Company Ltd
38 Russell Square, London WC1B 3QQ

www.britishmuseum.org

Susanne Küchler and Andrea Eimke have asserted the right to be identified
as authors of this work.

With the exception of the following, all photographs are © The Trustees of
the British Museum:

Pages 6, 12, 13 (left), 21 (centre), 30 (left), 31 (left and top right), 32, 33
(bottom left), 34, 46, 53 (left), 60, 67 (bottom), 74 (bottom), 81 (bottom),
92, 96, 97, 99, 103 and 108 © Andrea Eimke; page 87 © Ewan Smith;
page 88 © Phil Bender; page 113 © Andi Merkens.

Rights associated with the textiles and other artworks pictured in this
volume remain unaffected by their reproduction here.

A catalogue record for this book is available from the British Library

ISBN 978-0-7141-2580-0

Editing and project management by Sean Kingston Publishing Services
Typesetting by Sue Bushell

Printed and bound in Hong Kong by Printing Express Ltd

The papers used in this book are natural, renewable and recyclable
products and the manufacturing processes are expected to conform
to the environmental regulations of the country of origin.

Suffolk County Council	
07438909	
Askews	Jul-2009
746.4609	£25.00

Contents

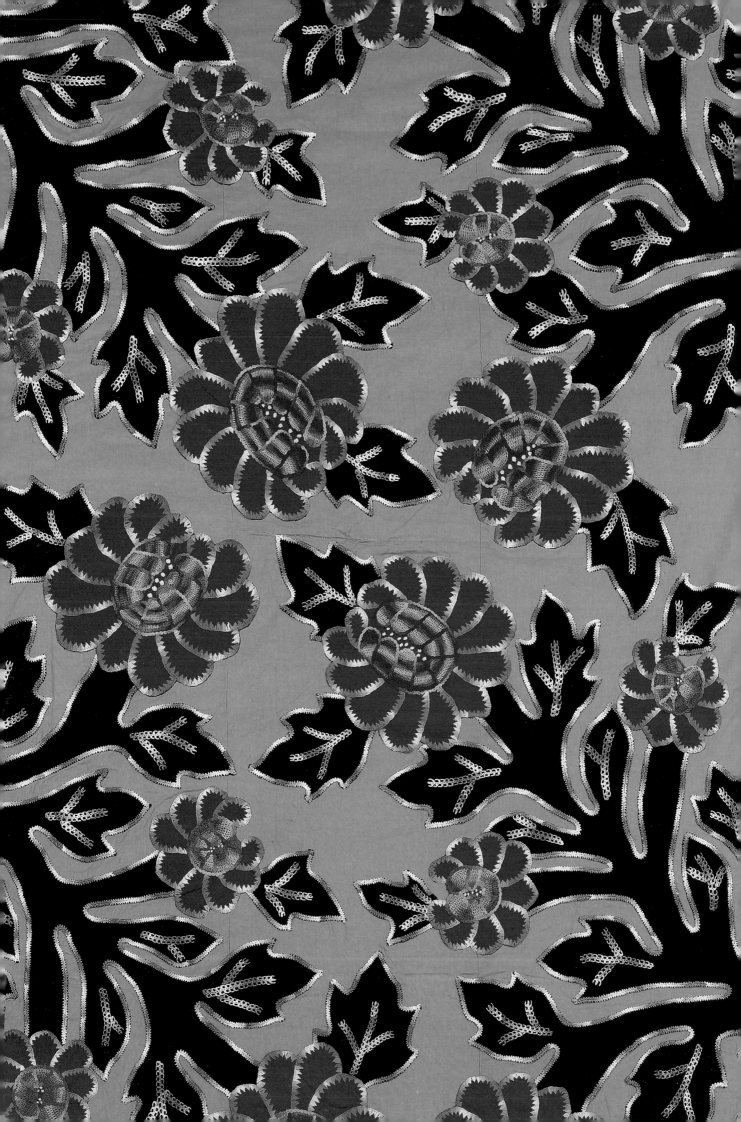

Acknowledgements

The research for this book was funded by the UK's Economic and Social Research Council as part of the 'Clothing the Pacific' project (ESRC number ROOO 23 91 98) hosted jointly by University College London, Goldsmiths College and the British Museum. The photography for this book was supported by a grant from the Getty Foundation.

Many people have helped in making this book, and we give our thanks to them all. Our warmest appreciation goes to the people of the Cook Islands, to whom this book is dedicated, and to the many women who invited us into their houses and greeted this project with open arms. Of these, we would like to highlight: Vereara Maeva, Tokerau Munro, Parau Taruia and Terangi Little on Rarotonga; Kimiora Samuel, Krudean Kareroa and Patricia Makike Glover on Mangaia; Patikura Jim and Nga Mokoroa on Atiu; Tuatai Koronui in Australia; and Emily Teariki Rakei and Taukura Porio in New Zealand. They showed great patience in teaching us about the art of *tivaivai*. Gill Vaiimene deserves a thank you for her much valued assistance on Mangaia, and Andrea is indebted to Armagh Karoa for co-ordinating interviews in Auckland.

We are grateful to Liz and Tata Crocombe of the Rarotongan Resort and Spa; Te Tika Mataiapo Dorice Reid of the Little Polynesian; and Ngakura Kautai, Tira Tararo and Tangata Edwards of Atiu for making their priceless *tivaivai* collections available for photography. We thank Jane Mason and the Cook Islands Museum and Library Society, Rarotonga, for permission to include historic photographs from their archives. Thank you also to the British Museum for making its artefacts available for reproduction, and for its overall facilitation of this publication; and to Sean Kingston for his editing. Tini Tivini and Teremoana George deserve Andrea's thanks for their help with translations and advice, and for their cheerful equanimity. Juergen Manske-Eimke receives fond appreciation for his help with logistics, his enthusiasm, encouragement and patience. Josephine and Isabell deserve Susanne's thanks for their invaluable help as companions and assistants on the many research trips, as does Ian for his patience and encouragement.

This book has benefited greatly from the critical commentary it has provoked at various stages of production. Special thanks go to Susanne's colleagues at UCL, who supported her in the often difficult juggling of teaching and research commitments, and who showed interest in her musings on the materiality of *tivaivai* and the story of its impact on culture and society in the Cook Islands. Particular gratitude goes to Nicholas Thomas, who originally suggested Cook Islands *tivaivai* as a topic for inclusion in the ESRC project, and who encouraged Susanne to take it up. His knowledge of Pacific art and history has been invaluable throughout. All remaining mistakes and omissions are entirely the authors'.

The main text is written by Susanne Küchler. Andrea Eimke shared her long-time expertise of living with *tivaivai*, composed the text boxes and took the majority of the photographs that illustrate this book.

Tivaivai tataura 'Chrysanthemum' designed and embroidered by Emily Teariki Rakei, Atiu; 209 x 262 cm (detail). Collected by Dr Susanne Küchler and presented to the British Museum in 2003. BM Oc,2003,02.6.

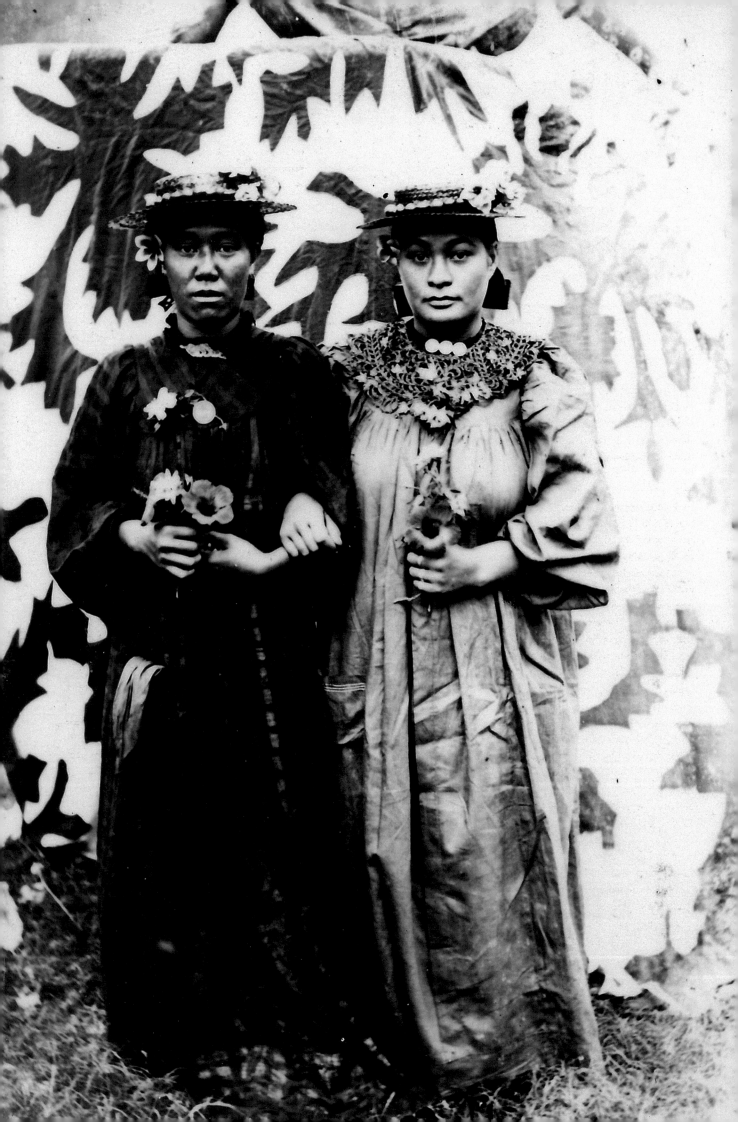

Chapter 1

A short history of patchwork in the South Pacific

Patchwork, the world over, evokes associations of warmth and nurture. Whether hand-stitched or machine-sewn, the patched-together pieces of cloth appear to make tangible the forgotten, the strange and the foreign in one swift move: making all pasts equally present. Of all the artefacts capable of reviving memories, patchwork illuminates in the simplest and most convincing manner the potential for material and the marks left by the virtuous application of technology to shape the mind. Made from discarded pieces of fabric, or from cloth bought specifically for cutting and assemblage, the weight, smell and look of the patched cloth brings a place and its people to life more concretely than any merely visual representation could. Patchwork is decidedly not about representation, but is rather about ways of being and thinking, 'mind work' we might say, and is about creating a strange analogy with what is 'social' about identity.

There are many ways to describe the associations provoked by patchwork, and those who have begun to be attentive to its role in society find it hard not to marvel at its capacity to pave a network hidden from the bright light of the everyday. There are the quilts made by Afro-American slaves to encode, in the stitched patterns, the pathways to freedom known evocatively as the 'underground railroad'. Quilts from the frontier of colonial America provide evidence of an underground economy run by female heads of households on the back of cloth-wealth. These precious fabrics were transformed into a material heritage to which proprietary rights could be extended and transmitted in ways that supported the independence of the household. Many who love the making of patchwork enthuse about the activity's generation of connections with loved ones, both those whose former clothes may be reworked into patterns that are evocative of loss and of a remembrance of times gone past, and those who will become the future recipients of the finished artefacts.

Patchwork connoisseurs speak about the capacity of patchwork to create and sustain relations that make up the meaningful and coherent thread that enables a life to be recounted as having a 'biography'. The absence of patchwork

may make a difference to the way biographical relations are managed; though the work put into patching pieces of cloth into myriads of patterns may be meaningful precisely when such relations cannot be taken for granted. Clusters of relations founded upon such work with cloth may emerge precisely, like other networks relating to the making and receiving of artworks that inspire the sharing of a common memory, when people do not 'feel at home in the world', but instead live in diaspora (intellectually, emotionally, politically, economically) (Belting 1995). The conditions under which identity may come to be considered a scarce resource to be husbanded are manifold, and we may suppose that they have only increased with the practice of labour migration and the growth of entire transnational societies that trace relations to increasingly remote and imaginary homelands. It is with the fragility of such relations in mind that we turn to the equally fragile patchworks made by one of the most scattered transnational peoples, whose 'homelands' are tiny islands in the middle of the Pacific Ocean. On these islands, living in a diasporic condition was required by the landscape long before the labour trade, travel and work permits increased the expansion of the community.

This study is set in the Cook Islands, an archipelago of fifteen tiny islands. The total land area of the country is only 240 square kilometres, while its exclusive economic area covers a maritime region of nearly 2 million square kilometres, stretching between Tonga and Samoa on the one hand, and French Polynesia on the other. Since 1901 the group has been included within the boundaries of New Zealand, and its people, who are culturally close relatives of the Maoris of New Zealand, are citizens of that country.

The country is broadly divided into a southern and a northern group, on account of the very different ecology and geology that dominate life on these islands. The islands of the southern group – Rarotonga (the main island), Aitutaki, Atiu, Mangaia, Manuae, Mauke, Mitiaro, Palmerston and Takutea – are, with three minor exceptions, of volcanic origin (and continue to uplift) and have fertile soils and lush vegetation. The exceptions are the small atolls of Palmerston and Manuae, and Takutea which is a sandy key. The southern group comprises about 90 per cent of the Cook Islands' total land area, with Rarotonga being the largest island with the highest elevation. The northern group consists of seven islands of coral formation, all being low-lying atolls with sparse

The photographer has come to town. Two women posing in front of a *tivaivai manu*. Photo courtesy of Cook Islands Museum and Library Society, Rarotonga.

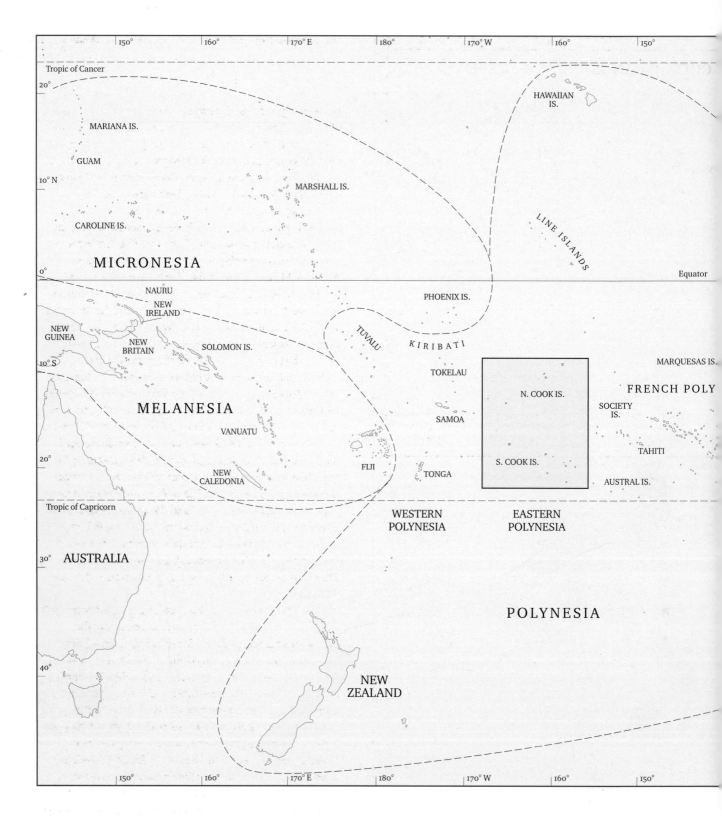

vegetation (mainly coconut and pandanus trees) and large lagoons. The closest outer island to Rarotonga is Mangaia (204 km distant), while furthest is Penrhyn (1,365 km away).

The islands' size, ecological diversity and distance from one another means that relations between them can only be maintained through the expense of effort, time and, today, money. While travel to even the neighbouring outer islands is still associated with a prolonged stay, to warrant the expenditure, it would have been even longer prior to Cook Islands' efficient inter-island airline, when waiting for the next

ship to arrive could take weeks, if not months. Moving away must always have been an enticing prospect, given the low density of population on most of the islands. Palmerston, in the southern group, is least populated, with just 63 people in 2006; only two islands, Rarotonga and Aitutaki, have more than 600 inhabitants.

The indigenous population, known as the Cook Islands Maori, is Polynesian and is closely related both to the indigenous populations of French Polynesia, some 1,100 km to the north-east, and to the New Zealand Maori, some 3,000

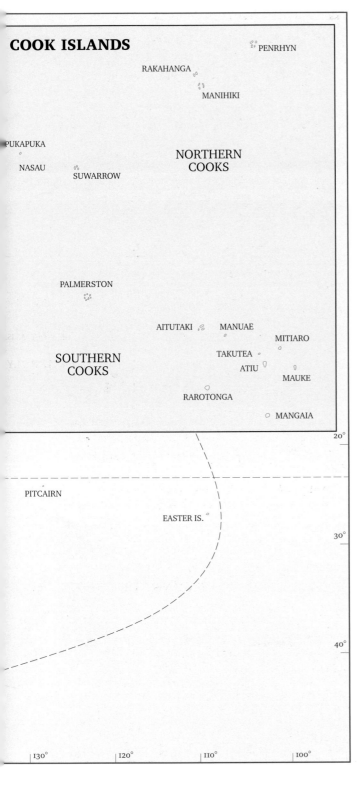

COOK ISLANDS

PENRHYN

RAKAHANGA

MANIHIKI

PUKAPUKA

NASAU   SUWARROW

NORTHERN
COOKS

PALMERSTON

AITUTAKI   MANUAE

MITIARO

TAKUTEA

SOUTHERN   ATIU
COOKS

MAUKE

RAROTONGA

MANGAIA

20°

PITCAIRN

EASTER IS.

30°

40°

130°   120°   110°   100°

wave of migration, admittedly in the context of a counter-flow of foreign workers to the touristic and administrative centres, was initiated by an economic reform programme of 1995–6,[1] when large numbers of Cook Islanders took permanent residency in New Zealand, Australia and the United States. Today, more than 50,000 Cook Islanders reside in New Zealand and an estimated 15,000 in Australia.

The islands are believed to have been settled from what is now French Polynesia, around AD 1200. The first European to come across the archipelago in 1596 was a Spaniard, Alvaro de Medana. Though the islands derive their name from the explorer Captain James Cook, he never actually set foot there. It is the arrival on Aitutaki in 1821 of the Reverend John Williams, from the London Missionary Society, together with two Society Islanders named Papeiha and Vahapata, that can be taken as the onset of contact, missionization and colonialism proper. The historical figure of the Tahitian convert Papeiha, who was the only member of the missionary party to stay in Rarotonga after John Williams left, and 'who accomplished in two years more than the English missionaries in Tahiti had in twenty' (Gilson 1980: 21), is emblematic of the way Eastern Polynesia was converted to Christianity by other Polynesians, in a transformation that was as swift as it was seemingly unremarked. Today about 70 per cent of the population still belong to the Cook Islands Christian Church. To prevent French expansion into the Pacific, the British declared the islands a protectorate in 1888, with the newly independent New Zealand annexing the territory in 1901. The country became self-governing in association with New Zealand in 1965.

No one quite knows at what point after contact patchwork started in the Cook Islands, but what is clear is that the practice of stitching pieces of cloth into large rectangular sheets spread across Eastern Polynesia almost as soon as fabric imported by Europeans became available. Waves of exploration, and the subsequent establishment of trading posts, had a profound impact on this part of the Pacific, notably on the islands of Hawaii soon after 1778, when Cook arrived. The earliest records of Pacific quilting are from Hawaii, where Russian and Chinese trading ships landed huge quantities of cloth and where missionary wives seized on the keen interest of Hawaiian women in working with cloth to instil new habits of body and mind (Colchester 2003). Clothing the Pacific was anticipated as a huge endeavour on the part of the London Missionary Society, which began its operations in Eastern Polynesia. The zeal with which Polynesians adopted the introduced cloth for manifold purposes may have inspired the American Transcendentalist Henry David Thoreau to write in 1854, 'I say, beware the enterprises that require new clothes, and not rather a new wearer of clothes.' (Thoreau 1971[1854]: 23). To the missionaries, clothes, so readily adopted, almost appeared a distraction from a proper concern with the immaterial. But

km south-west. At the time of the first European contact in 1821 there were approximately 18,000 people on the islands, a number which declined sharply to 8,000 at the end of the nineteenth century. Today, there are again around 18,000 inhabitants, although the census of 2006 shows only 11,800 permanent residents, a drop of nearly 3,000 people from 2001 (*Cook Islands Times Weekly*, 30 Sept. 2007). There has long been a net outflow of Cook Islanders. In 2006, the number leaving for overseas was 969; the highest annual exodus was that of 2000, when 1,429 left the islands. The most substantial

across Eastern Polynesia, once the technology of sewing took hold and woven cloth became more readily available, clothing the dead became the most effective way of servicing relations with the spiritual domain. Hitherto, it had been exclusively men that controlled the items of ritual and sacrificial exchange with which relations to the realm of the unseen were managed (Küchler and Were 2005). As Polynesian women took charge of fabricating the new items of ritual exchange, which across Eastern Polynesia took the form of patchwork, so they also became the keenest converts to Christianity. Quilt-making and the economy of exchange it supported prospered under the tutelage of the Church. New – and yet, in a profoundly Polynesian way of approaching the future by looking to the past, so decidedly *not* new – it is perhaps no surprise that stitching quickly became the dominant emblem of emerging national identities.

Clothing and the imposition of Christianity

It has been argued that pre-colonial and pre-Christian ideas surrounding fibre as a material technology of wrapping the body may have laid the grounds for the Polynesian interest in cloth and patchwork (Thomas 1999). We know that the clothing of the Pacific happened in no time at all, and while the speed

and degree of fervour with which introduced cloth and clothing were adopted by Pacific Islanders may have varied from place to place, women in the Cook Islands appear to have seized upon both the imported material and ways of making as soon as they became known to them.

Cotton reached the Cook Islands as early as the 1840s through whaling ships, which were cruising in the Pacific in large numbers. In 1843 a report notes that 'no less than 35 vessels called here (in Aitutaki) during the last 12 months' (Beaglehole 1957: 69).[2] Shortly before 1850, the American whaling fleet numbered 680 ships, of which all but 40 were stationed in the Pacific; over 100 a year are known to have called at Rarotonga for supplies. Cotton was grown on Rarotonga in the 1850s, barely 25 years after missionaries first set foot on the island, and the local printing of cotton cloth produced in the Cook Islands soon followed. The ethnographer Ernest Beaglehole (1967: 87) describes for the island of Aitutaki the swift take up of cloth in the Cook Islands:

> In 1827 men wore *maro* loin-cloth, women a short petticoat of bark cloth, children up to ten or twelve years nothing. In 1857 the men wore coats, waistcoats, shirts, trousers, hats; some, shoes and stockings. The women wore an inner garment of bark cloth, on top a long flowing robe; on their heads bonnets of finely wrought plait trimmed with gay ribbons. All children were 'decently'

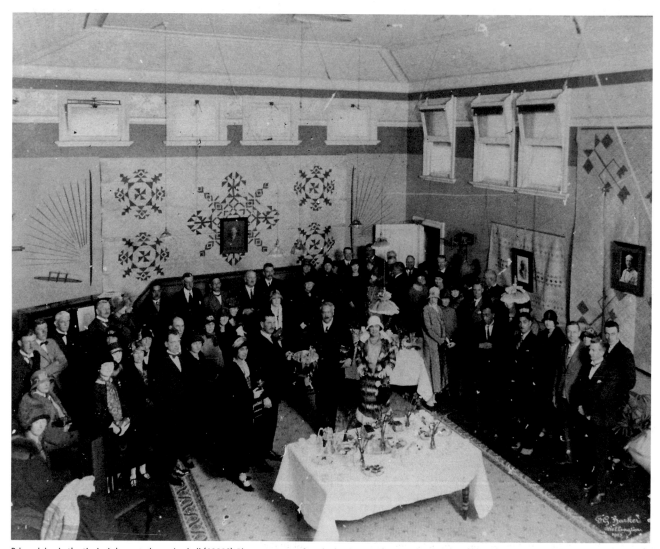

Prize-giving in the *tivaivai*-decorated meeting hall (1929?). Photo courtesy of Cook Islands Museum and Library Society, Rarotonga.

clothed. Although this clothing was all completely unsuitable for the climate the missionaries were thoroughly satisfied that the changed circumstances were met and proper.

Woven cloth did not just change outer appearances, it also impacted on modes of deportment and attitudes to cleanliness and work, as such clothing must have constricted movement and made work in gardens a difficult undertaking. Again, Beaglehole (1957: 87–8), referring to the autobiography of the missionary Buzacott (Sunderland and Buzacott 1866: 236–46), describes vividly the changes in the domestic economy: 'In 1827 dwellings were of customary style, described by Buzacott as "mere wigwams". By 1857, every family was said to have a good cottage for itself… Chairs, tables, sofas and beds were the furniture in all houses.'

The historian Richard Gilson (1980) points out the increasing status of women, who filled almost all chiefly positions from the 1850s onwards, with the self-declared Queen of Rarotonga visiting New Zealand in the 1890s. The first account of patchwork is in a 1906 novel by Clement L. Wragge. Cloth stores are one measure of the popularity of patchwork: already established in Rarotonga in the mid 1850s, on the outer islands they were set up by the 1880s. On Atiu, the first cloth store was established by a Chinese merchant who arrived from Tahiti. Today, cloth stores such as Vonnia's of Rarotonga and Atiu import bales of plain coloured cloth called 'Azlin', recognized by its loose weave and heavy weight, as well as sewing and embroidery thread in all the colours of the rainbow.

Missionaries saw the adoption of European clothing as a sign of conversion, yet the manner in which cloth and clothing were imposed and received varied greatly from place to place, partly reflecting distinct histories of colonization and missionization, but also representing differences in localized conceptions of cloth and its transformative potential across the Pacific. Richard Eves (1996), for example, argues in his discussion of Methodist missions in New Britain, Papua New Guinea, that the introduction of clothing was only 'skin deep', resulting in merely superficial changes that could be observed in the transformation of the imported clothing itself, such as the unravelling of knitwear and its reworking into openwork designs, yet which did not lead to lasting transformations at the societal level. Graeme Were (2005), in his account of the nineteenth-century trading of printed cotton in this part of the western Pacific, shows how calico was not simply worn around the waist or tied around the neck like an apron, but was transformed by tearing it into strips that adorned heads rather than bodies. By contrast, Nicholas Thomas argues that for Eastern Polynesia introduced cloth constituted a kind of technology towards a new being in the world through its impact on deportment and attitudes to cleanliness and domesticity, including time-keeping (Thomas 1999; cf. Küchler and Were 2005 and Colchester 2003).

To this day, dress, and in particular women's dress, is marked by the diverse histories of clothing in the region. The Anglicans introduced blouses and skirts for women, a style that remained characteristic of women's dress in North Vanuatu and the parts of Papua New Guinea led by the Anglican Church. The Presbyterians, on the other hand,

especially in southern Vanuatu, introduced dresses known as 'Mother Hubbards' that were a simplified version of European dress in the nineteenth century (Bolton 2003 and 2007). The long sleeves and long skirts of these dresses made it difficult for women to maintain physical labour, giving rise to gender differences and new sentiments of domesticity that were more pronounced than those of regions where cloth was radically altered to suit pre-existing expectations. The 'Mother Hubbard' dress has been adapted to numerous local styles, and is commonly decorated with binding, ribbons and lace or, at the very least, with pleats folded into the fabric – features of dress which have become a symbol of a rural and island-based conservatism that has become fashionable again in the Cook Islands today.

The Hawaiian *holokû*, which originated in 1820 as an adaptation of an American day gown, shows how woven cloth facilitated the translation of pre-existing ideas of status into material forms that have outlived their colonial past. First adopted by Hawaiian queens as a means of achieving a layered look in dressing the body, an arduous achievement with barkcloth, it was subsequently worn by Hawaiian women of chiefly status as day wear. The *holokû* is now formal wear for ritual and festive events celebrating local identity. As in Hawaii, the other island nations of Eastern Polynesia had long associated the wearing of upper-body garments with status. Museum collections document the importance of so-called *tiputa*, ponchos made of barkcloth, which were worn widely across French Polynesia and the Cook Islands during the early nineteenth century. The donning of layered upper-body garments by leading figures aboard ships, and by high-ranking Polynesians themselves, was a factor in persuading islanders of the power of stitched clothing and Christianity, and that the adoption of both were ways of harnessing it. A further example of the transformation of European-introduced cotton cloth and ways of dressing into visual icons of Pacific modernity is the Samoan two-piece dress (*puletasi*), modelled on the twin set.

Clothing, however, was not all that was inspired by the arrival of the new cloth material and the new technologies of sewing and cutting that allowed for its innovative transformation. Home furnishings – such as pillow covers, throws, bedspreads and wall hangings – have been produced by women all over the Pacific, but nowhere as notably as in Eastern Polynesia, where women have been sewing huge elaborate quilts for over a century. Measured in units of the time devoted to their stitching, and valued accordingly, patchwork sewings are gifts of the highest value and are rarely on display. Instead they are commonly stored in treasure trunks until they are returned to the maker, often on the occasion of the creator's death, when the deceased person is wrapped in them before being lowered into the grave. The association of cloth with binding the dead and the living to each other is reflected in the visual imagery of these quilts, which use symmetry in creating strikingly abstract designs, and yet also contain the concrete in their references to flowers and plants whose acquisition punctuates the biography of the quilt-makers.

Tivaivai as ceremonial cloth

Tivavai have long been precious to Cook Islands women, though few households in early twentieth-century Atiu would have been able to afford the quantity and quality of cloth required to make them. They may have started as fashionable home embellishments, featuring reproductions of flowers that only grew in the far-away countries the missionaries came from. Yet the focus of the women who introduced the art of making them would primarily have been on the transmission of technical skills and Victorian principles of domesticity.

But remember: a Cook Islander's life was previously wrapped in tapa cloth. Here you had a valuable textile item, the pride of one's household. Why not use it in ways tapa, which the women were no longer making, was once used? And so it was that *tivaivai*, instead, began enveloping a Cook Islander's life.

Tivaivai mark as special the areas in which important stages in an islander's life are celebrated. Whether a celebration is held at home, or the event is considered important enough to hire a community or church hall, *tivaivai* cover all the walls and give the room a joyful and colourful look.

The seat on which someone honoured by a special celebration is to rest will be covered with an important *tivaivai*. Often friends and neighbours come and borrow our peacock chair, which with its high backrest provides a most suitable 'throne'. A special litter is prepared for dignitaries' visits to our island. This is carried by strong men, dressed in costumes fashioned from dried banana leaves, which symbolically prevent them from touching the dignitary and being contaminated by his *tapu*. The seat is covered with the finest *tivaivai*. Because our airport is quite far from the villages and the Boys Brigade Ground, where important events take place, important visitors are taken by car, in which the seats are, naturally, covered with *tivaivai*.

Sometimes even the ground over which the honoured dignitary walks gets covered with *tivaivai*. When Fr. Christiano, a newly ordained Catholic priest and son of the neighbouring island of Mauke, first visited Atiu, a path of the most beautiful *tivaivai* was laid out on the grass to cover the ground of the Catholic church from the gate to the stairs. Of course, his seat in church was also covered with a *tivaivai*.

Tivaivai are part of a young couple's dowry and are piled up before the newly-weds. They form part of the gifts given during

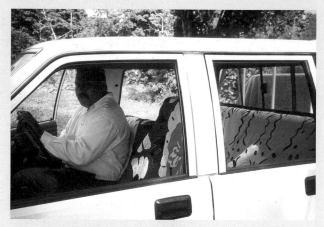
Tivaivai cover the vehicle in which dignitaries are collected at Atiu airport.

special celebrations such as hair-cutting ceremonies or twenty-first birthdays. They are presented by visitors to their hosts as *kura* (an object of veneration, a precious gift, formerly of red feathers; hence the name *kura*, which means 'red'), and hosts give them to departing visitors as farewell presents, to show their appreciation, love and emotional bond. They are also presented in exchange for gifts received, on familial and official levels. In the 1990s the Cook Islands received a patrol boat from the Australian Government, for which they expressed their appreciation with three embroidered *tivaivai*.

Special *tivaivai* form part of a mother's wedding gift to her daughter. These are often destined to become the funerary shroud for the daughter's husband. When a family member dies, they will be laid out in state in the family's home for a final farewell, their bed covered with *tivaivai*. Those *tivaivai* touching the body will be wrapped around it when it comes to be placed in the coffin. Further *tivaivai* may be piled up on top of the coffin once it has been lowered into the grave, before finally being covered with earth.

Even death does not conclude a person's contact with *tivaivai*. A year or so after a funeral, the grave will be adorned with a headstone. Close relatives and friends will bring *tivaivai* to cover the commemorative stone. In turn, each of them is then called upon to come forward and lift one of these precious cloths, layer after layer, until the headstone is finally revealed. It is only then that the deceased can finally rest in peace, fondly remembered by those who loved them and continue to love their memory.

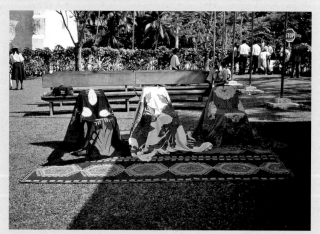
Three people will receive an official decoration. Their seats are covered with *tivaivai*.

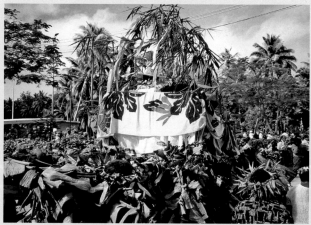
Dignitaries are carried, by strong men dressed in dried banana leaves, on a litter covered with *tivaivai*.

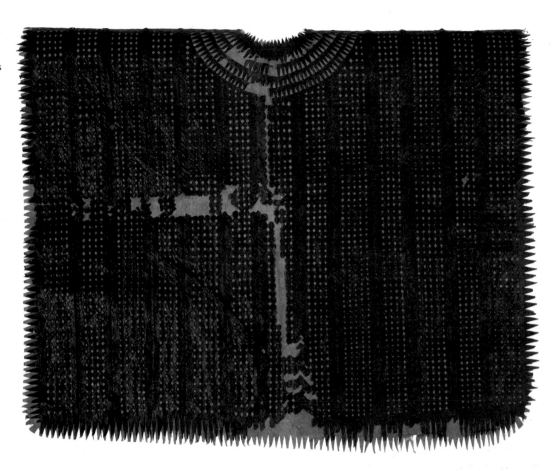

Tiputa mangaia,
c.122 x 107 cm. Collected
by Revd William Wyatt Gill.
Presented by Sir A.W. Franks
in 1876. British Museum
Oc,9953.

From fibre to fabric

The historical figure of the Tahitian convert Papeiha, who was the only one of the missionary party to remain in Rarotonga after John Williams left, helps us to consider the manner in which this part of Eastern Polynesia was converted to Christianity (Gilson 1980: 21): less by inculcation and force, than by suggesting new technology that would improve existing material means of controlling the relation between the visible, the realm of light, and the invisible, the realm of darkness, from where *mana*, or ancestral power, emanates. The proximity of these realms meant that places and persons could be temporarily *tapu,* a condition of heightened sanctity brought about through the dissipation of *mana* to the surface of bodies or landscapes via openings such as bodily orifices or waterholes. To enable persons and places to return to ordinary life, de-sanctification procedures such as washing or wrapping the body or place were at least as important as sacrifices to the gods, as without them objects, places or persons remained dangerous and potentially lethal because of the volatile impact of uncontrolled life-force or *mana.*

We know from ethnographic collections that Tahitians had long utilized fine cloth inscribed with thinly drawn motifs, created from the beaten bark of the paper mulberry tree, as a removable bodily wrap. As Tahitians, who had already embraced Christianity with its attendant material culture of cloth and of sewing, converted the Cook Islands and Samoa, such upper-body garments, known as a 'ponchos' in ethnographic collections, were adopted there also as symbols of high status by chiefly ranks. For the Cook Islands,

collections of ponchos held in the British Museum show that mourning cloaks worn at funerals had a distinctive upper layer which was stained black and covered in diamond-shaped cut-outs arranged in parallel rows along the entire length and width of the garment. In addition to the cloaks, special tapa (barkcloth) cloths were made by men of particular tribes to be used as offering to the gods.

We know about the mid-nineteenth century of Polynesia partly through archival sources of missionary origin, but also through ethnographic collections, such as the textile collections of the British Museum, that contain many examples of barkcloth that have been elaborately stained in bright yellow turmeric or deep reds and covered in hand-drawn lines or cut-outs. Such textile collections were soon supplemented by so-called 'idols', the assumed material link between people and their deities, which were carved and cordage-bound sculptures, called 'staff-gods', whose binding and untying punctuated the most important ceremonies for controlling the presence of *mana* on the islands in the Cooks and in French Polynesia. Cook Islands staff-gods were elaborately carved 'sticks', sometimes more than three metres in length, along which vertically arranged faces or figures were duplicated down the shaft. In other staff-gods, miniature versions protruded like outgrowths from the central figure, carved in the round, whose body appeared to have served as a treasure chest. Between ceremonies (when the staff-gods were kept in the tribal ceremonial ground or *marae,* itself delineated by a stone boundary) these protrusions appeared

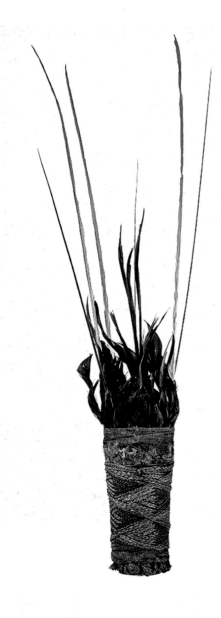

Above: Large staff-god wrapped with painted barkcloth, from Rarotonga. Length 396 cm. Ex collection London Missionary Society. Acquired by the missionary Revd John Williams in 1827. British Museum Oc,1978,Q.845.

Left: Small god image of coconut fibre, barkcloth, feathers, fibre (from Atiu?) Height 52.3 cm. British Museum Oc,LMS 53.

to be covered with cordage, bound into flower-shaped feather holders, and layers of patterned white tapa cloth.

It is important to note that the power attributed to these staff-gods rested not so much on their representational properties, as in their capacity to harness *mana* through the periodic containment and evocation of their image. In its form of a manifold, a multiple, this 'hidden' image expressed its potential to sustain what we colloquially call the 'fertility' of land and of people, but which may have extended to life in general. Today, the unwrapped images of the gods, usually in their singular form, present some of the most important artefacts of the past transformed into souvenirs, speaking eloquently about a new ritual, even seasonal, economy that now sustains life in the Cooks in ways that are not without past parallels.

These staff-gods reached museum collections as they were given up in the process of conversion to Christianity, and became, like the ponchos made of barkcloth, obsolete agents of ritual technology, superseded by what came to be called *tivaivai* (Jones 1973; Akana 1986; Hammond 1986; Pogglioli 1988; Rongokea 2001; Shaw 1996). Much of

what the new skill of sewing and the availability of fabric would have meant to nineteenth-century inhabitants of the Cook Islands can now only be inferred. Sewing is still one of the main skills that brings women together, under the leadership of pastors' wives, and it is clear from the historical reconstruction undertaken by Richard Gilson that women had come to occupy most of the chiefly *ariki* titles in the late 1900s, with support from the Church. In her dealings with British officials, and in visits to New Zealand, the Rarotongan female chief of the most prosperous district, a certain Makea Takau, came to be known as Queen of Rarotonga, and inspired the take-up of European parasols and 'colourful frocks' as essential parts of Cook Island fashion (Gilson 1980: 50–1).

Perhaps it is because the European observers and missionaries who populated the Cook Islands between 1850 and the present day are predominantly male that we know so little about the role women played in shaping and integrating an increasingly diasporic society. Beaglehole (1957: 89) describes in disparaging terms the effects of missionary influence and the imposition of new laws that undermined the authority of the chiefs in favour of policemen and judges, whose main qualification for administering the new laws 'was skill at scripture elucidation and preaching'. The principal chiefs had lost initiative and independence in society. Younger men who would have been likely to succeed their elders appear to have been lost in droves, as a result of whaling fleets replenishing the large numbers of runaway sailors by taking young islanders aboard, who often left willingly, and against the wishes of their families, with an eye to the large sums of money promised for work on the ships. Many, though, never returned, frozen to death in Arctic waters; or eventually returned home severely ill. Others simply deserted their vessels at Californian ports, so as to prospect in the newly opened goldfields, and often never returned to the islands (Beaglehole 1957: 72). American whaling was at its peak in the years 1830–60, and between the years of 1843 and 1862 the average whaler lost almost two-thirds of its crew from desertions and discharges.

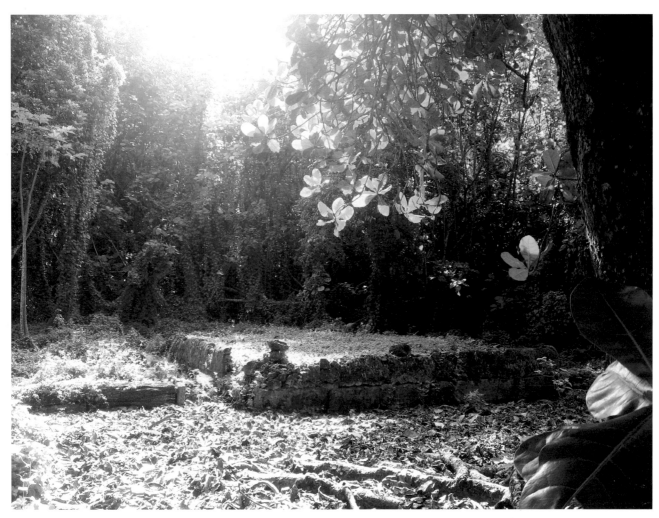

Stalactites and stone slab structures on Atiu's Marae Orongo.

Yet the hiring of young islanders as sailors was not the only reason for the rapid decline of able-bodied men in the Cook Islands during the late 1900s. The greatest worry to both Cook Islands missions and to islanders was Peruvian slave vessels, which frequented the islands between 1860 and 1867. About 1860, agricultural developments in Europe and the rising value of cotton had suddenly made Peruvian guano valuable – it was used as fertilizer in the new cotton plantations of Peru, Fiji, Tahiti and Queensland – and Peru experienced an agricultural boom in cotton and sugar. In a search for cheap labour, Peruvian ships sailed the central and southern Pacific, taking some 10,000 Pacific Islanders to work in the mines, on the plantations and among the guano deposits (Beaglehole 1957: 95). All but a fraction died in Peru, leaving many Pacific islands depopulated until Britain and France joined in bringing pressure on Peru to ban 'blackbirding' slave raids by 1864.

One can note from these events the irony of history, which allowed the heightened value of cotton production in Europe and America to inflict devastation on islands whose people, for very different reasons, had a very similar desire for cotton, making it into the symbol of modernity par excellence. Given the number of traders stationed on the main islands, cotton was never in short supply. Beaglehole (1957: 119) reports letters exchanged between missionaries stationed in the Cook Islands in 1889 which note an increasing presence of trading firms in Rarotonga: examples included a New Zealand firm, a branch of a German firm, the Société Commerciale de l'Océanie from Tahiti and an American importing house. Given the evident role of cotton in trade relations among newcomers and residents, it is surprising that no attempt has been made in the extensive publications about the Cook Islands to reflect on the difference made by the production and circulation of sewing to political institutions in the islands.

Political and religious institutions in the Cook Islands

Ron Crocombe, in his book *Land Tenure in the Cook Islands* (1964), describes the effects of the environmental and demographic conditions prevailing in the diverse islands of the Cooks upon a system of social organization that was based on a pattern brought from Eastern Polynesia. The concepts described below capture ideal types of a social organization whose status increasingly comes under question today, as tendencies towards intra-tribal marriage and intra-tribal adoption, brought about by pressures on demography

(continues page 12)

Experts in exile

Tara Maui

'Mama would be so pleased to know that you remember her!' May Maui's excited voice beams at me through the phone. She sounds almost the same as the last time I saw her, on the occasion of a brief interview with her mother in 1991. When I started taking the first notes for what would later become this book, Mama Tara was already terminally ill; she passed away a few weeks later on 29 March 1991.

Two-thirds of a century earlier, on 20 July 1924, Tara Maui had been born on Atiu. As a very young girl she may still have been able to learn from Mama Teakarua Moni, the woman considered to have introduced the first sewing groups to Atiu. When she married her husband on the island of Mangaia in 1947, at the age of 23, she was already a well-known *ta'unga* and had cut many *tivaivai* for women on Atiu and Mangaia. Her most beautiful patterns include highly intricate designs such as 'Hanging Lantern' ('*Mori Tautau*'), 'Butterflies' ('*Pepe*'), 'Crowns' ('*Korona*'), 'Crowns and Pineapple', 'Fans' ('*Tairiri*'), 'Kiss Me Quick' and many others.

When I asked May whether she could bring along a *tivaivai* of her mother's, she had to admit regretfully that she had given hers away as a gift at a family function. After some persuasion, she was able to borrow one belonging to her brother. We went to the Domain park in Auckland, and sat on the lawn for a long chat and a photo session. These pictures show how special Mama Tara's patterns were, different from those of any other *ta'unga* on Atiu. The craftsmanship in this *tivaivai*, which she sewed herself, is immaculate. Evenly sized and spaced stitches throughout, with tidy finishing on the back, were her trademark.

May remembers that Mama Tara was also skilled in making tapa and weaving pandanus mats and hats. I recall her sewing hats: joining long braids of pandanus on her old treadle sewing machine. Mama Tara was an outstanding seamstress, who could look at a dress, or the photograph of one, and then get the scissors, cut the fabric, and sew one just like it. Consequently, she had to sew all the village uniforms for the *tere tere* Pentecost celebrations.

'We'd come home from school and the house would be full of bags with fabric for the uniforms already cut up,' May remembers. 'She'd spend two days and nights sewing and we'd have to help. Or she'd have cut a *tivaivai* and spread it out on the floor for tacking. She'd get mad with us, telling us to be careful not to step on the fabric, and we'd complain, "but Mom this is our house where we live! We need to be able to step somewhere to get in!"'

But Mama Tara would not accept their complaints, and instead would make them sit down and help her with the tacking or other sewing jobs. Like many young women (myself included), May was not keen to learn how to sew. Back then, helping her mother seemed a tedious chore.

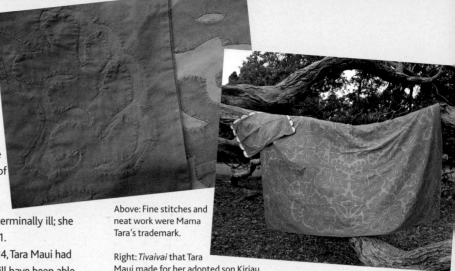

Above: Fine stitches and neat work were Mama Tara's trademark.

Right: *Tivaivai* that Tara Maui made for her adopted son Kiriau.

Some time ago May, now a mother of two, joined a group of Manihiki women in her Auckland neighbourhood who gather for sewing *tivaivai*. 'Those elderly women, some of whom had to be taught how to sew, were quite surprised at the skill with which I handled my sewing jobs, until the *ta'unga* told them that I had grown up as a *ta'unga*'s daughter,' May proudly tells me.

Mama Tara was always very modest about her talent. When women put money in the folded fabrics they sent to her for cutting into a pattern, she used to return it, refusing to accept payment. Nor did she want to keep the food women brought her as payment. They would have to sneak into the house, quietly leave it on the table and quickly disappear, so that she could not argue with them.

'A short time ago, for a wedding, I decided to sew waistcoats for my three boys all on my own, and it was as if my mom was right behind me, telling me how to do it,' she smiles. 'When I picked up Mom's *tivaivai* from my brother's house and drove down here, I kept looking at it sitting next to me in the car. It felt as if Mom was there, feeling pleased that her work was receiving the recognition it deserved.'

Very few of Tara Maui's works have remained on Atiu. The women for whom they had been cut gave most of them away as treasured heirlooms to family members that now reside overseas. Though this shows that her patterns were very popular, and were therefore selected as special gifts, it saddens me to think that her work has almost disappeared without trace from our island. But she left her memories in our hearts, and I feel privileged to write these lines in honour of her life's work.

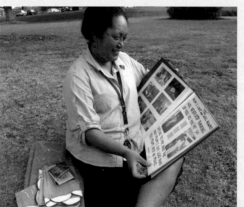

The pages May lovingly prepared in this photo album are all that remain to remind her of her mother.

Taukura Porio (Mama Tapu)

Atiu Place No. 1 in Mangere, Auckland. We leave the car parked in front of the huge house and enter. Passing a tidy office, well equipped with computer, printer and everything people of today need to organize work, I admire the glass case on the opposite wall, which is filled with trophies of all sizes: first, second and third prizes, for all sorts of competitions the members of our Atiu community in Auckland have succeeded in. The hall in front of us is so huge that it dwarfs the eight women that have gathered around a large

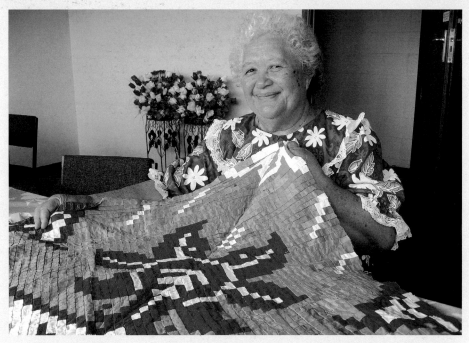

Taukura Porio, a *ta'unga* in Auckland, with a pattern sample from the 1940s.

table. And then I see Mama Tapu: she hasn't changed a bit! I immediately recognize her by the halo of white curls that surround her friendly face. In front of her is a piece of fabric the size of a cushion-cover, on which she has just finished drawing a cluster of frangipani blossoms. The woman sitting opposite has now to trace these onto a sheet of transparent plastic, in order to copy the reversed design to the first cushion-cover's pair. We hug and kiss, and Mama Tapu introduces me to the seven women next to her. I do not know any of them, and nor do they know me; most have been living in Auckland much longer than I have lived on their home island of Atiu. But I know many of their family members: brothers and sisters and cousins, and some have heard of me and my husband.

On my right side sits a petite octogenarian, skilfully stitching a red snowflake pattern onto its yellow background. Despite her advanced age, she does not wear glasses, but her stitches are neat and tiny! From her repeated questions, I gather that she finds it hard to believe that I, the *papa'a* (white foreigner), have lived on her home island for twenty-three years.

'Let's go to the other room,' invites Mama Tapu, and one of the *mamas* follows us with two cups of coffee and a plate of biscuits on a tray.

Mama Tapu is the daughter of one of Atiu's first Members of Parliament, Papa Vainerere Tangatapoto. When we visited Atiu for the first time in 1982, he flew back home on the same plane, in those days a five-seater Sky Master. His obvious love for, and engaged description of, his island won our hearts for Atiu even before we had set foot on its shores.

Papa Vainerere came from a family of tailors. He and his brothers all knew how to sew men's suits, Boys Brigade uniforms and shirts. Their mother sewed the female garments, long frilly dresses made of floral cotton prints. Young Tapu grew up in the family home surrounded by fabrics and the rhythmic sound of treadle machines. Her mother, Aketua Vainerere, had died aged 26, when Tapu was only three years old. Soon she was given easy sewing tasks to help her grandmother.

When she was seven years old, her mother's first cousin took her to Rarotonga, where she lived in the Takamoa Mission compound until her return to Atiu in 1946. She had learned how to use a sewing machine and started to sew her first mosaic piecework *tivaivai* at the age of twelve. She was also a skilful weaver of pandanus mats, baskets and hats. In fact, when my mother visited us on Atiu in the mid 1980s, Mama Tapu gave her a beautifully plaited *kikau*

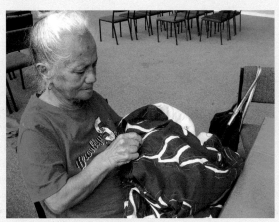

Never too old for sewing *tivaivai*.

(coconut-leaf fibre) hat, whose colourful hatband, composed of minute shells, has survived to this day.

In 1949 a small group of eager students at Atiu school were chosen to become teachers. Tapu was one of them, her subject being, of course, home economics and sewing. In the 1950s Mama Tapu got married and had her first son, Tauu Porio, the only child who still lives on Atiu today. The Women's Institute was established and Mama Tapu joined a sewing group. She discovered her talent for designing *tivaivai* patterns. Her first *tivaivai manu*, a lily pattern, accompanied her fourth son to his grave. The *tivaivai taorei* that she brought with her to the interview is her own first *taorei* pattern, a rose. She shows me a triangular fragment from one her grandmother's *tivaivai*, made in the 1940s, that she has kept as a family heirloom. Despite her sixteen children, she has continued teaching and sewing .

By 1987, when her daughter started studying at Auckland Polytechnic, most of the older children had left home, and the family moved to New Zealand. Though no longer teaching, she kept busy creating outstanding *tivaivai* patterns. Her design skills were in high demand for all sorts of family and official celebrations, and her patterns became famous in the Auckland Cook Islands community. In September 1999, she formed the sewing group whose *ta'unga* she remains today. When she shows me a folder in which she has gathered meticulous records and photographs of her own artistic career and of the group's events, I am amazed and full of admiration. The group had their first exhibition outside Auckland in Wellington in 2003 and has been invited to Christchurch for workshop events. They have also been the subject of an anthropological thesis and have been written about in textile magazines and a recently published book about the Atiu community in Auckland. She proudly smiles when I tell her that in 2005 the C.I. Tivaivai Association exhibited two of her *tivaivai* (which her group had cut and sewn for Revd and Mrs Tangaroa Uea, then based on Atiu) at the 11th Patchwork Meeting in Sainte-Marie-aux-Mines, France.

resulting from out-migration, have begun to elevate new forms of association (such as sewing bees organized around the affairs of the Church) to becoming economically and politically important entities. Today, the connections through which ideas, money and influence are circulated are figured in a decidedly horizontal manner, conjoining makers and receivers through the memory of the pattern. And the contemporary assemblage of pattern draws a visual analogy with modes of incorporation that have the potential to transcend the hierarchical and exclusionary nature of social relations demanded by the political organization of the time of European contact. As the vertically organized social network based on genealogy and hereditary rank is today bisected by horizontal, conterminous relations and groupings, which cut across existing boundaries dictated by birth and social position by means of elevating the political and economic relevance of ties of friendship and adoption, the 'traditional' concepts will be described in the past tense.

The most encompassing social unit, which effectively controlled the ownership and transmission of land, as well as the transfer of hereditary titles and of knowledge, was the tribe, known also as *vaka* or canoe. The tribe was composed of all those who traced their descent from persons who had travelled to the island in mythical times on the same canoe. The leadership of the tribe rested in the position of *ariki*, to which incumbents were ascribed on account of their direct descent, ideally in the male line, from founding ancestors of some 48 or more generations ago. These ancestors were venerated as god-like, and were deemed to have bestowed their supernatural powers upon their descendants. Crocombe describes how in the course of the twentieth century alone, the number of titles in the *ariki* class within a tribe varied from several title-holders to none at all. Such titles are also known to have been downgraded to the head of a landholding section of a tribe, the *mataiapo*, for reasons possibly connected with the need to defend land-rights that became increasingly localized as tribes began to marry within their

Times of change

The re-establishment of self-government in 1965 resulted in a new pride in Cook Islands ethnicity and in a revival of traditional values. It was then that the 'outer islands' (the islands other than Rarotonga) began playing an important role. Their distance, isolation and relative autonomy from the centre of development in Rarotonga had resulted in less constant exposure to Western influences, and thus the greater preservation of the indigenous language, tradition and material culture.

When houses were built, fishing canoes carved, food crops planted or crafts practised, the young ones would be invited to be present and to watch. They would also be asked to run errands and do menial tasks. Some of them, of course, considered that bothersome, so they stayed away. Others remained, because they were curious and keen to learn. Eventually their interest would be rewarded and they were given the privilege of helping the experts.

Today this pattern is changing, even in the outer islands. Children are no longer around all day, because they have to go to school. Knowledge is acquired from books. But schoolbooks contain little about building houses from natural materials,

carving canoes or the secrets of traditional crafts. The school curriculum used to include woodwork and crafts, but that is no longer so. In any case, organized education cannot be compared with hands-on experience of village traditions.

The modern influences of media and migration bring different needs and requirements. Because the old experts cannot find young enthusiasts who want to acquire and practise and eventually pass on their wisdom and skills, traditional knowledge is gradually being lost. Many have died, taking their secrets with them to the grave, while a new generation is growing up in a different cultural environment with values, languages and traditions imported from other lands.

There is a new generation of Cook Islanders growing up in foreign countries. More than twice as many Cook Islanders live overseas as in their own country: most in New Zealand and Australia. Even though many of them still live in the Cook Islands communities that have formed in certain towns, traditions cannot be adhered to as firmly as back in the islands. Life is faster, more demanding and more expensive. Many islanders were sent overseas to help with the family budget; money has to be saved and sent home.

The 'new' Cook Islanders have not had the opportunity to sit with their elders and watch canoes being carved or *tivaivai* sewn. Few can relate to the time involved in making such a precious cover. When ceremonies require the giving of 'cloth', it is easier to buy synthetic factory-produced bed quilts, woven acrylic blankets in bright colours or printed sheets. For the folks back home in the islands this may even be appreciated, as these items are harder to get there, the finance may not be so readily available and duvets, blankets and sheets can be used in daily life rather than being stored in a glory box. Some appreciate the value of a *tivaivai* from a different perspective though. They have learnt that *tivaivai* can have a monetary value, and can be sold in times of financial need.

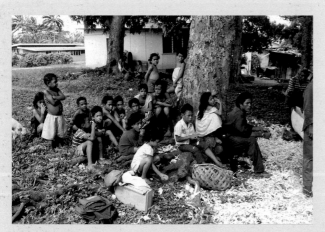

A canoe is being built; the village kids have come to watch.

own boundaries. The *vaka* was a decision-making unit, and during pre-contact times it organized and compensated others for warfare. Each *vaka* owned one or two *marae*, stone-walled enclosures that were the centre of ritual activity.

Islands were divided into tribal districts called *tapere*, each of which was headed by a named *mataiapo*. The occupants of a *tapere* were known collectively as *matakeinanga*, and, significantly, this included all residents, whether their connection was based on birth, marriage or otherwise, an organizational feature which facilitated the incorporation of outsiders in the early years of contact, from deserters of ships to missionaries. To single out those related by descent (reckoned in relation to an accepted member) within such resident groups, a local descent group identified itself as *ngati*. Descent was patrilineal, although a man could inherit his title from his mother. Descent groups branched over time into numerous sub-groups, usually as a result of marriage within the group, and these in turn were headed by a leader, called *rangatira*.

A person belonged to the descent group of both parents, although children were assigned a primary affiliation at birth, a status which was confirmed through naming. Most children were assigned to the resident group, although adoption by the other descent groups or even outside the tribe was common. The predominant pattern of adoption was for the child of a woman who had married out to be sent to its mother's father or brother in her natal descent group.

Significant in Ron Crocombe's account of traditional social organization is the fact that individuals retained the right to potential membership in all the other resident descent groups to which one could trace connection, either through adoption, birth or marriage in his or her own present generation or in previous remembered generations (Crocombe 1964: 30). The extended family group (*kopu tangata*) was the most encompassing social unit, cutting across tribal boundaries and residential affiliations.

Arguably, one reason for the success of the Cook Islands Christian Church was that it based its principles of

So they give away their precious gifts, often for much less than the monetary value really applicable to such a work of art.

Luckily, there are others that have taken their traditions and their love for this craft with them. Sewing groups – in which Cook Islands women are now busy designing, cutting and sewing *tivaivai* – have formed overseas. Exhibitions have been held in many museums and galleries around the world, and artists have been able to display their own and their family members' works to an international audience. Overseas quilt-makers are eager to learn new techniques, and where possible join Cook Islands women in sewing groups. These interchanges provoke new ideas, as do publications such as quilting magazines.

Some of the foreign ideas and innovations have been brought back home to the islands, and have gradually become apparent in the local *mamas' tivaivai*. Crochet is one example. Now many squares with applied or embroidered motifs are joined together with lace, bought by some, but proudly crocheted by those who have acquired the new skills. In the 1980s some *tivaivai* makers discovered that, before applying a motif to the background with a sewing machine, they could crochet around the flowers and leaves (rather than embroidering the outline), thus producing some eye-catching results more quickly. A Rarotongan sewing group recently exhibited patchwork *tivaivai* that used printed fabric, instead of material of a solid colour, inspiring their fellow artists with an entirely new look. These innovative developments show that *tivaivai* making is still very much a living tradition, even though fewer pieces are made and young women are no longer prepared to patiently sit and sew as before.

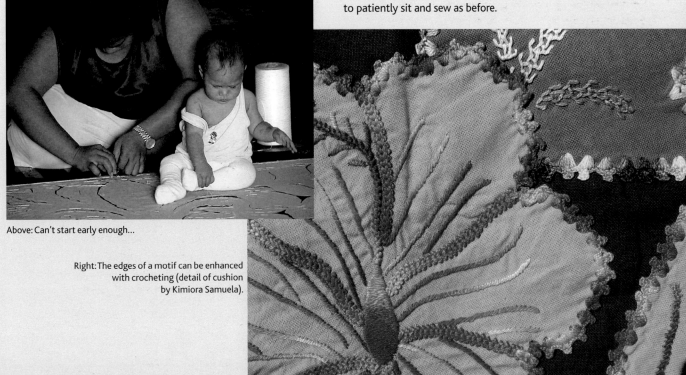

Above: Can't start early enough...

Right: The edges of a motif can be enhanced with crocheting (detail of cushion by Kimiora Samuela).

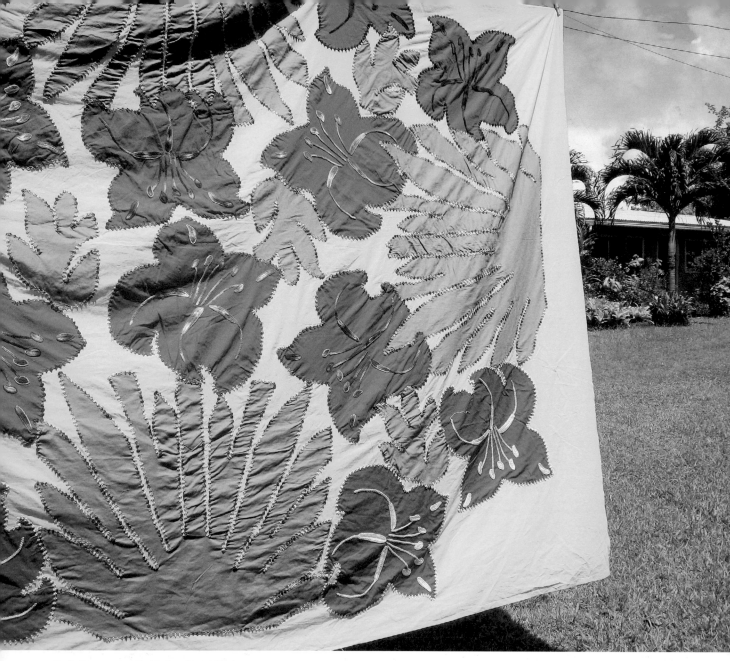

Tivaivai tataura owned by Ngakura Kautai (Atiu).

membership on existing patterns of social relations, whose tripartite division offered the combinatory possibility for an infinitely expandable group. Just as *vaka*, *tapere* and *ngati* affiliation was based on spatial and temporal principles of belonging, so the level of the household also had its own tripartite division, with a localized *ngati* or descent group being subdivided into clusters of houses that occupied what were known as *kiato* segments, which in turn were divided into *kainga* (households headed by an elder known as *metua*) and *puna* (households comprising nuclear families). Affiliation to a Church equally follows the principle of residence, which can be conceptualized as scaled according to principles of encompassment from local, to island to inter-island or transnational membership.

We will meet this principle of tripartition, created through scaling and multiplication, across all the expressive arts of the Cook Islands, including dance, song, poetry and stitched patterns on the surfaces of quilts. At work here is a tripartite structure that by the process of analogy can be moved from one domain of life and from one medium to another, reproducing in fractal fashion, and connecting everything that is touched by it into an infinitely expandable nexus of social relations.

The mutually encompassing groupings which made up the political institutions of the Cook Islands at contact were image-based polities, defining their identity in relation to equally layered images that took the form of the afore-mentioned staff-gods, carved from ironwood on the inside in fractal, self-replicated representations, and wrapped in layers of cordage and barkcloth. The main ritual, described by the anthropologist Alain Babadzan (1986; 2003) for Tahiti, centred on the dramatic unwrapping of the figure, which released the spirits of the dead, and the rebinding of the figure to refocus the powers of the spirits on the land demarcated by the stone-walled *marae*. *Paiatua* or 'the gathering of the gods' was also a central ritual event for the Cook Islands, and collections appear to indicate the use of different scales in distinguishing images that belonged to groups of differing encompassment.

Alongside the shift of the centres of religious and political organization from the stone-walled *marae*, perhaps the most dramatic change effected by the Church has been the supplanting of the agentive factor of descent from named and imaged ancestors to individual persons, who as *ta'unga* began to control the transmission of patterns in song, dance and fabric that connected everyone on account of their inherently recognizable fractal template. Arguably, it is this transposition of image-based agency onto persons which is endangering the remnants of 'traditional' Cook Island social organization today, as out-migration and depopulation have broken the cycle of transmission within the homeland, transposing its workings onto increasingly cut-off diaspora populations. As the resource of knowledge dries up in the homelands, it is questionable how long the connective capacity of pattern-making can be sustained as a lived principle in the diaspora.

Why quilts?

The future, and indeed the past, is shaped by practical inventions that subtly alter the infrastructure of life. The development of needle, thread and cloth would have been one such innovation capable of changing the pattern of everyday life; changing the way people walk and hold themselves, changing attitudes to the body and to cleanliness, to the gendering of work and social space (as, for example, long dresses made it impossible for women to maintain work in the gardens, and demanded they spend the daytime inside the house). Perhaps its most important change was in attitudes to time, which could now be seen to be invested in seemingly durable things that could be treasured and stored, exchanged for services and future returns, or converted into other things of equal value.

Quilts may appear to us as a leftover from an age when people still *had* time – time to do things that literally soak up time and make us cherish the product of work as evidence of time well spent. Indeed, the only object that commonly remains part of the everyday and that clearly speaks of the time which has been invested in the making of it, is food. From the Cook Islands perspective, such a likeness between sewing and cooking makes a lot of sense, as both are essential in managing the important events in one's life and have the same effects: to extend, when given away, the social relations which a person encompasses; and, surprisingly, to create memories anchored in multi-sensory perceptions. For large ceremonies, food in the form of sweet potato and taro tubers is prepared in large earth ovens dug into the ground, with the raw and peeled produce being placed into them in rectangular baskets plaited from the leaves of the coconut palm: the smells emanating from the cooked food charm the senses in the same way as the bright and colourful patterns of a *tivaivai* ensnare the eye.

The geometric and mathematical construction of Cook Islands *tivaivai*, so eloquently exhibited in their symmetrical and fractal composition, reflect in large measure the fabric's indexical and symbolic value as an agent of temporal reckoning. *Tivaivai* patterns enable the recall of certain events associated with key moments in women's lives and encode ways of categorizing social relations that unfold in the vicinity of the quilts. For patchwork in the Cook Islands is not made primarily for use: quilts are taken out of their treasure boxes merely once a year for 'tutaka', an inspection of the house by officials from the department of health that goes back to colonial times. Instead, patchwork is sewn to be given away as treasure, or *taonga*. The concept of *taonga* is shared across many nations of Eastern Polynesia.

Made famous by the anthropologist Marcel Mauss (1923/67) in his work *The Gift*, we know that a *taonga* can be material or immaterial, 'dead' or 'living'. In fact, such distinctions are not made (Henare *et al.* 2007: 47–8): a *taonga* can be a piece of whale bone, a local plant or piece of shell, women or children, a woven cloak, but also specific knowledge and skills or the language of the Maori. Today, important *taonga* are products of technology as well as products which international law is only beginning to recognize as property, such as products of gene technology. A *taonga* is all important to the Maori, not just because is it

Embroidered embrace

Mata Teamoke and I, as directors of the Atiu Fibre Arts Studio, travelled to San Francisco in 1992 to attend the opening of a *tivaivai* exhibition at the Folk Art Museum in Fort Mason. When we arrived at the museum one morning, the receptionist told us that a young man had asked after us the previous day. He would drop by in the afternoon, she said. We wondered who that could be, because we knew nobody in San Francisco.

It is afternoon. Many interested viewers have passed through the exhibition this day and we are proud to be given the opportunity to acquaint the visitors with our island country's textile art.

When a tall man in his twenties enters the room, we know instantly that he must be the caller. His features betray him as a Cook Islander. He comes straight up to Mata and introduces himself with a big, friendly smile. 'I'm originally from Aitutaki,' he tells us.

'I must be the only Cook Islander who lives in San Francisco. I'm an actor and have lived here for two years now. People in the States don't know about the Cooks. When I saw the news coverage about the *tivaivai* exhibition I was so excited, I had to come straight away! Good on you to show Cook Islands textile art over here! It is so cool to see all those beautiful *tivaivai*. They remind me of the islands. It's a good feeling, but it makes me homesick at the same time. I am lonely here sometimes,' he admits with a sad smile.

'I brought you something to look at,' he continues, and the expression on his face lights up again as he pulls from his large shoulder bag a beautiful *tivaivai*, richly embroidered.

'This was the farewell gift from my grandmother when I left Aitutaki to go to the States. She raised me and made this especially for me. You know, it somehow reminds me of her. I get it out often, especially when I'm feeling homesick. It's like she's with me then, and puts her arms around me to comfort me. San Francisco can get so cold and grey, especially in winter, and then this *tivaivai* is like the warm colours of home.'

a carrier of life-force, but because it *is* life-force or *hau* (New Zealand Maori), a concept which underscored Marcel Mauss's account of the power of the gift.

Mauss's contribution to what we know today as 'thing theory' consisted in his exposition on the identification of person and thing, which he found in the early ethnographic documents from the Pacific. The inspiration for his critical thoughts about the nature of things was derived from his encounter with the so called 'ancestor figures' and ceremonial artefacts collected in large quantities by missionaries and explorers during the early exploration and colonization of the Pacific, which embody such personalized powers that they are still believed efficacious in recalling past owners today.

Foundational for his studies about the relation between person and thing was a letter written by a Maori chief from New Zealand, named Tamati Ranapiri, to the English ethnologist Elsdon Best. Significantly, in this letter Ranapiri equated the notion of *taonga* or object of value with the notion of *hau* which was translated as the life-force or 'soul' of the gift, a connection which led Mauss to develop his theory of the social obligation to return a thing that is given. Mauss famously argued that when a *taonga* is given away, it carries with it the soul or *hau* of the donor. It is the unbroken connection between donor and the thing given that forces the recipient to return this 'vital' agent of exchange.

> The taonga that I received for these taonga (which came from you) must be returned to you. It would not be fair (tika) on my part to keep these taonga for myself, whether they were desirable (rawe) or undesirable (kino). I must give them to you because *they are a hau of the taonga that you gave me*. If I keep this other taonga for myself, serious harm may befall me, even death. This is the nature of the hau, the hau of the personal property, the hau of the taonga. (Mauss 1990: 11)

The quilt as personal property and as gift exposes us to ideas which have become critical to theoretical models aimed at explaining how societies in the Pacific have turned themselves into thriving transnational societies that sustain economic flows between the many places of the diaspora and the homelands (Evans 2001). Cook Islands *tivaivai* serve as an excellent example of how the original creator of a *taonga* may remain embodied in the object, as both pattern and stitch bear recognizable personal traits that commemorate a particular personal 'handling' of the skill of sewing and stitching. Not only can women in the Cook Islands spot with ease the maker of a *tivaivai*, but they are also full of stories of how

Tivaivai tataura sewn and owned by Te Tika Mataiapo Dorice Reid.

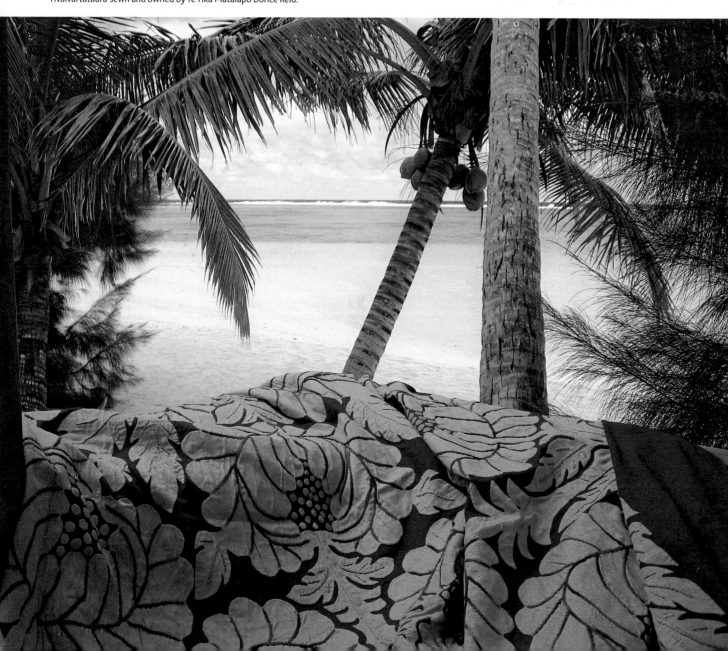

such personal traits are transmitted across generations, from mother to daughter or granddaughter.

Even though all patchwork is made to be given away as a gift, not all quilts circulate equally rapidly. Already in his original work on the gift, Marcel Mauss had highlighted a basic distinction underlying Polynesian *taonga*, between those that are 'movable' (*'meuble'*) and those that 'do not move' (*'immeuble'*), but he had not understood what this had to do with the inalienable nature of the gift, enforcing its ultimate return to the person whose *taonga* it is. The anthropologist Annette Weiner (1992) elaborated upon this distinction with her work on fibre arts in the Pacific, showing that artefacts that circulate with enhanced velocity maintain the value of those that are kept as treasures in one place. Her theory of inalienable possessions explicated this notion of 'keeping while giving', which showed gift exchange to be central to a model of cultural property whose capacity to withstand the pressures of a commodity-based economy had until then not been understood.

What we may want to take from this for an understanding of patchwork in the Pacific is the importance of a relation that is deemed to exist between persons and things, allowing certain things to stand in for or to act as a substitute for persons. Recent anthropological theory has used the understanding we now have of the dynamics of gift exchange to reflect upon the nature of the kind of artefact that can serve as *taonga* and thus act as substitute for a person in exchange. This theory develops from a critical reflection upon the obviation of distinctions which we hold for granted in the Enlightenment traditions of Euro-America: distinctions between wholes and parts, between singular and multiple, between things that are dead and things that are alive, and fundamentally of course between person and thing. Drawing upon an earlier theory of a notion of fractal personhood, in which a person is seen to be composed of multiply authored and partible substances, the anthropologist Alfred Gell has expanded on the material and technical aspects of artefacts that complement such a notion of personhood, enabling such artefacts to exude the kind of agency that is normally ascribed to persons (Strathern 1988; Gell 1992; 1998). His book *Art and Agency* explores the logic of such a formal relation, which allows mere things to become cognitively and emotionally 'sticky', in that they attach themselves to persons on the grounds of a perceived likeness or formal connection.

Gell's ruminations on the general conditions for an anthropological theory of art that explains the nature and constitution of art-like situations in generic terms may make more sense to us when it is related back to the Pacific and to the Cook Islands quilts, which as *taonga* should match his description, if his theory is correct. With reference to the Pacific, art-like things, which create social relations in the manner of a gift, are said to be typically of fractal composition and consist of singular motifs that are multiplied through scaling and additive construction (see Henare *et al.* 2007). Here, technical virtuosity, which guarantees the efficaciousness of the artefact and its capacity to act in an art-like manner, is argued to consist in the handling of geometrical and mathematical problems, but also to depend upon the right kind of material that allows for partibility to be a recognizable principle of visual expression.

Cook Islands patchwork may indeed be seen to be iconic of a notion of fractality, which it shares as a formal principle of composition with the other quilt traditions of Maori-speaking populations in Hawaii and Tahiti. Azlin, the type of cotton fabric available for patchwork construction today, is, however, a material that is not very easy to rip, one reason perhaps why real patchwork made of thousands of tiny pieces is dying out today, and why after Azlin began to be used the individual squares became larger: one just couldn't rip it into 1-inch squares! The material used until the 1970s was muslin, a much finer all-cotton fabric which was lighter in weight and easy to rip. Though Azlin eventually replaced all the imported muslin, its heavier weight made it much more usable for appliqué, giving the unbacked *tivaivai* more body through weight. Apart from the constraints of the material upon construction, the essential work of the composition of patchwork patterns is of multiplication and ensuring commensurability, as an assemblage of one core motif is repeated in a symmetrical fashion in scaled proportion to a preconceived whole.

Cook Islands *tivaivai* will take us on a journey of discovery, exploring how connections are formed and maintained through simple thread and cotton fabric among a people who embrace the condition of living in diaspora as a part of their history and landscape. Though nostalgia is part of what is created, this journey should also make us reflect that sometimes the simplest and most traditional of things may be those best suited to dealing with the conditions of the future.

Notes

1. The reform program cut down the superfluous number of public servants that had been put in office through political favouritism. After the lay-off of approximately sixty per cent of public servants, the public service regained its productivity, but it had devastating economic consequences for the country.

2. Early publications on the Cook Islands include: Ida Lee, *Captain Bligh's Second Voyage to the South Sea* (London, 1920); John Williams, *A Narrative of Missionary Experience* (London, 1938); William Gill, *Gems from the Coral Islands* (London, 1856); J.P. Sunderland and A. Buzacott, *Mission Life in the Islands of the Pacific* (London, 1866); J. Pakoti, 'First Inhabitants of Aitutaki', *Journal of the Polynesian Society* (1895 (4): 59–70); F.J. Moss, 'Maori Polity of the Islands of Rarotonga', *Journal of the Polynesian Society* (1894, 3: 20–6); Te Rangi Hiroa (P.H. Buck), *Material Culture of the Cook Islands* (Aitutaki) (New Plymouth, 1927); H.L. Shapiro and P.H. Buck, *Physical Characters of the Cook Islanders*, Memoirs, P.B. Bishop Museum (1936, vol. 22, no 1.); R.W. Gosset, 'Notes on the Discovery of Rarotonga', *Australian Geographer* (1940: 3, 4–15); P.H. Buck, *Arts and Crafts of the Cook Islands*, B.P. Bishop Museum Bulletin 179 (1944); P.H. Buck, *Introduction to Polynesian Anthropology*, B.P. Bishop Museum Bulletin 187 (1945: 92–6).

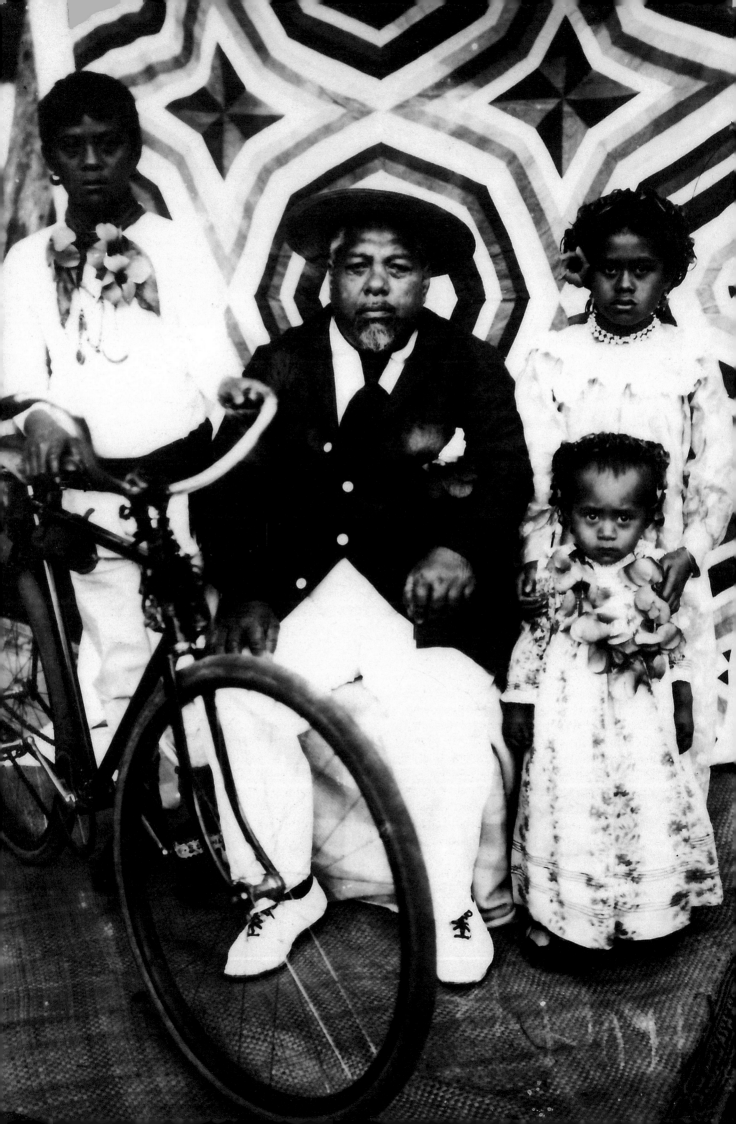

Chapter 2

Threads of life
The material connections

Migration has been a fact of life for Cook Islanders from at least the time of the earliest historical records. Massive migration, assisted by extensive networks of marriage relationships, has created a transnational society extending far beyond the shores of the homelands. This study is about these homelands, which for Cook Islanders remain spiritual homes (however imaginary they may be to some) and the 'sources of knowledge' (*tumu korero*). A considerable proportion of Cook Islands income comes in the form of remittances from relatives living as migrant workers in New Zealand (to which Cook Islanders have free access as New Zealand citizens) or further afield in the Pacific metropolises, in America or in Europe. Belying the romantic anthropological stereotype of an isolated local world, living standards are no different to those of New Zealand, with imported food, videos, television and the mod-cons of everyday life being found in every house.

Despite the transnational character of the corporate kin group, family ties with those who reside in the homelands are strong and flourishing, with family members travelling back and forth a great deal for annual or family ceremonies, such as Christmas, New Year, weddings or hair-cuttings or, most importantly, unveiling ceremonies, which conclude the commemoration of the dead. Even those who reside in the homelands, for however long, consider themselves to be part of this great wave of migration. As women are the most active migrants, they have great stories to tell of travels and residencies elsewhere, or are hatching plans for the future. Visits with the scouts to the Philippines, with the sewing bee on tour in Hawaii, or exchange visits of church groups to neighbouring local islands or to far away Australia are among the frequently mentioned occasions for travelling abroad, alongside the perennial trips to daughters or sons working in New Zealand or beyond. A poem written by Michael Tavioni for his granddaughter, Chrystalla Vaimana Ari'I Tefana-Neophitou, bears witness to the cosmopolitan life led by Cook Islanders for nearly 200 years:

Native of 2020

She is…
Fair skin
Green brown eyes
Slightly blond or red hair
Narrower nose
Longer neck
Pleasant looking
Thinner lips
An inevitable fruit cake

Local of 2020
Her gene bank includes
Rarotonga
Atiu
Mangaia
Tahiti
Aotearoa
German
English
Irish
Chinese
Spanish
Greek-Cypriot
Scottish

(Michael Tavioni in Mason 2003: 247)

Atiu: A Case Study

An obelisk-shaped monument stands on the Boys Brigade Ground in the heart of Atiu. It is five metres high, two metres wide and sits on a slab constructed of local limestone. Erected in the memory of Paulo Ngamaru Ariki, the Atiuan chief who assisted in the construction of the Cook Islands Christian Church nearby in the 1950s, the islanders call the monument *Te Pito* or 'the navel' and claim that it is situated at the exact centre point of the island.

When travelling around the island, this measure of certainty in its topology comes across in the almost perfect positioning of roads and houses, which radiate out from the centre in a star-like fashion. This arrangement of living 'on the top' is a product of twentieth-century planning; in former times the settlements were scattered in and around the foothill area near the coast.

Atiu is the third largest island of the southern group and forms a triangle known as Ngaputoru with the islands of Mauke and Mitiaro. Formed of a raised reef and a volcanic core, Atiu is a high island rising along a series of flat-topped ridges to 65 metres in the centre. It has an area of about 27

(continues page 23)

Tivaivai and bicycles: family wealth displayed for posterity (c.1902–7?). Photo courtesy of Cook Islands Museum and Library Society, Rarotonga.

A walk through Atiu

Atiu does not resemble the south-seas cliché.

The island emerged out of the ocean as a volcano, and gradually a coral reef formed around its edge. Erosion flattened the volcano, and the reef rose out of the water and petrified. A new reef formed beyond the edge of the old, and now almost entirely surrounds the island, protecting it from the destructive forces of the ocean. The small part (less than one kilometre) of Atiu's coastline outside the reef's shelter, Te Pari, offers an awe-inspiring sight. Standing on the razor-sharp fossilized coral reef, your eyes plunge into the blue of the deep waters below, broken by the white crests of powerful waves, whose pounding you can feel under your feet, like a heartbeat.

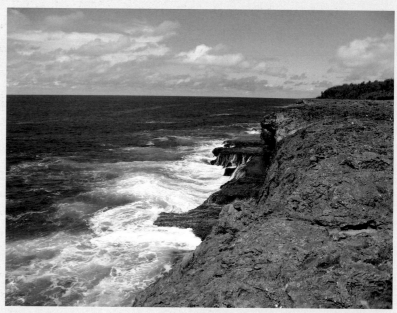

Raised cliffs on the north-east side of Atiu's coast.

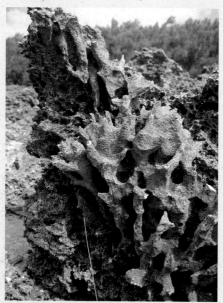

Razor-sharp fossilized coral rocks.

The islanders call the band of fossilized coral *makatea*. It provides the basis for a lush rainforest and is difficult to traverse. However, one can still find ancient paths, painstakingly constructed by filling the gaps between the sharp-edged rocks with coral slabs, evidently made to enable easier access from the shore to important gathering places.

The beauty of the *makatea* has lured me on many walks across the island. The coastal road is flanked by the magnificent giant *utu* (*Barringtonia speciosa*) trees. The late afternoon sun paints golden highlights onto their large leaves, which provide a strange contrast to the dainty blossoms that are about to reveal their magenta-tipped white filaments and golden anthers for only one sweet-scented night. The branches that cross over the road like an archway are crowned with the lanceolate leaves of the

kōta'a (birdsnest fern, *Asplenium australasicum*), along with its companions, the lacy *turei'āua* (*Davallia solida*) and the *tūroutou* (lobeless sword-fern). The *ngōtare* (kingfisher, *Todirhamphus tuta*) announces my arrival with loud protest; its steel-blue and white colours flash through the foliage as it escapes my intrusion into its green world.

This lush rainforest continues to provide many vital resources, such as food, timber and other building materials. The dense growth on the seaward side opens. I look down onto Oro Varu, the small beach at which Captain Cook's vessels once landed. It was not far from here that the first Polynesian settlers set foot on the island. In my mind's eye I can see the giant *vaka* (voyaging canoe) and hear the

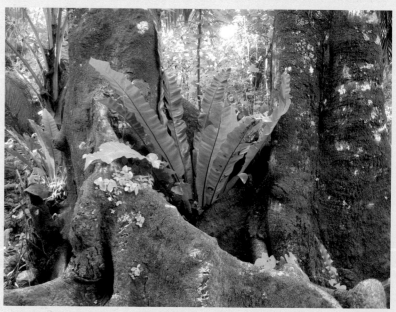

Birdsnest ferns grow on roots and branches of ancient trees.

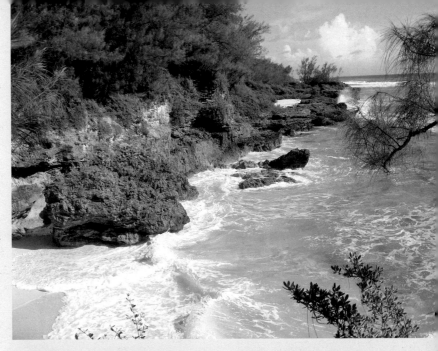

Oro Varu, the beach where Captain Cook's vessel once landed.

encouraging 'ooish, ooish' that accompanies the pulling of strong hands and arms as the catamaran is hoisted across the reef.

They once arrived on a double-hulled canoe, men, women and children, tired after a long voyage across the fierce waters of the South Pacific, and relieved to have finally sighted land where they might be able to settle and build a new future for themselves. Neither the reasons for leaving their homeland, nor which homeland it was, are known to us today. Throughout Polynesia, and in every local language (which themselves are remarkably similar), the origin of each island's settlers is known as 'Avaiki. The suggestion of common roots across the Polynesian Triangle is strong.

The settlers' journey was well prepared, with cargo that included animals and the most essential plants, to provide them with food, clothing and shelter during their journey and in their new life ashore. The crew likely incorporated experts in spiritual, medicinal and practical knowledge, whose solid foundation of wisdom would help their tribesmen and women create a new future.

The islanders never developed a written alphabet; they trained their memorial skills instead. Chants and songs bear witness to events that were and still are important to these nomads of the sea, the rhythms helping their commitment to memory. Many chants remind today's people of the adventures of the ancient seafarers during their long journeys across the sea.

Before the sun sets over the sea at Oro Varu, whose name and location suggest that this may have been the area where the souls of the dead were believed to leave the island to return to 'Avaiki, this land of dreams where all Polynesians once originated, I resume my walk back home. Between the *makatea* and the central plateau is a ring of low land (*raro-enua*). Soon, to the left of the main road, I will see an open area in whose swampy soil people used to grow taro (*Colocasia esculenta*), the starchy root that provides the most important part of everyday Polynesian diet. Like so many inhabitants of this island, all but one of the taro patches have gone, weeds taking over the fertile soil. On this island fresh water is not plentiful, and only here in the lowlands can trickly streams and waterholes be found. On the other side of the island they gather to form a small lake (Te Roto) that is connected to the sea by channels through the *makatea*. Its waters are slightly brackish. It is home to some wild ducks (*mokora-taetaevo*) and the delicious eel (*tuna*, Pacific shortfin eel). The bottom area of the slopes that rise up to the centre of the island provides fertile volcanic soil that lends itself well to vegetables and fruit trees.

While I climb up the steep road, I ponder the island's history. It makes sense that in pre-European times the islanders' dwellings were scattered at the base of the hills, close to their sources of fresh water (*puna vai*), their

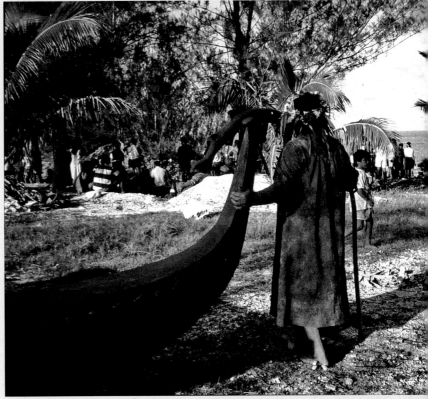

Memories of times gone past: Atiu's voyaging canoe is launched in 1992 to sail off to Rarotonga's South Pacific Culture Festival.

A fresh-water stream in the lowlands.

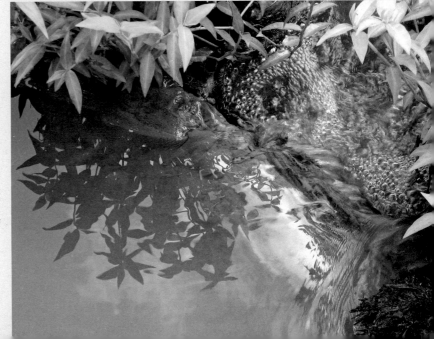

taro patches, their fruit trees and other important plant material. In a smart move to gain better control over his subjects, high chief Rongomatane relocated the people's settlements to the centre of the island, around the first Christian church, which had been built during his reign. To this day, Atiu's five tiny villages have remained arranged in a star shape that radiates from a centre on the island's summit.

Thinking of my host island's past has distracted me from the natural beauty that surrounds me. The main road's ascent is strenuous. The humming of the large

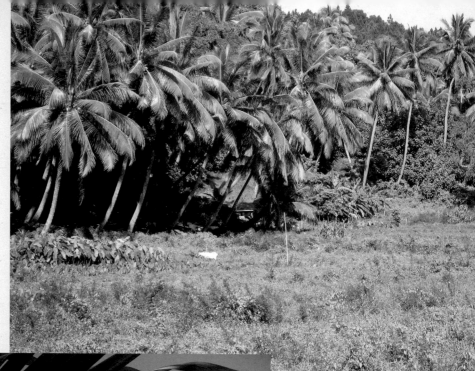

Raro enua, the lowlands that surround the island's centre.

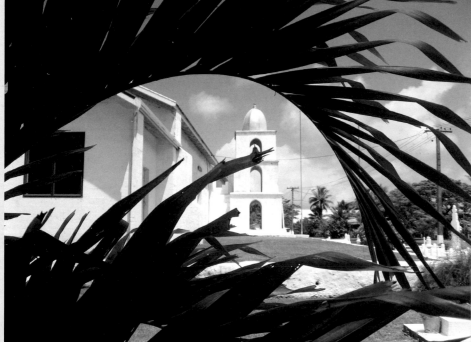

The CICC church building on Atiu stands out because of its bell tower.

diesel generators that provide Atiu with power tells me that I will soon reach the village I live in, Teenui. Our house lies on its fringe. The crimson sunset sky gives the dirt path an eerie light. The pigs that the neighbours keep in pens or tethered on ropes have gone to sleep for the day. The fading daylight takes the bright colours out of the plastic flowers that decorate the many graves in the family graveyard next to

our house. It is hard to read the inscription on the headstone of the first one I get to, but I almost know it by heart: 'Ina-i-te-Roe [...] died 27th August 1985, aged 107 years'. Few of her 15 children, 70 grandchildren, 70 great-grandchildren and 30 great-great-grandchildren still live on Atiu. There are graves of people I never met, some are so old that nobody remembers who lies buried there; and there are graves of good friends whose proximity is a comfort. Next to these is the new grave of my father, whose passing in 2006 I still grieve. This is now the soil where my roots lie. The bright lights of our yellow house wake me from my gloomy thoughts. I have reached home.

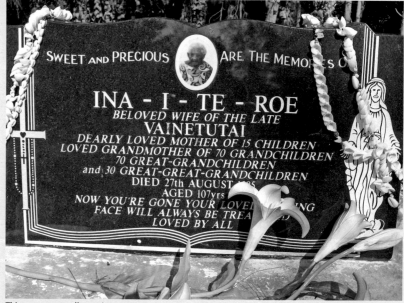

This gravestone tells a unique story.

square kilometres; it is drop-shaped and is approximately 7 km long and 5 km wide. An uplifted coral reef, the 'makatea', almost completely encircles the island. Rugged, rocky and in places impossible to cross, this band of makatea was once a rich source of supplies for Atiu people (Tanga 1984). There is no lagoon. Gaps in the reef connect the sea with deep underground caverns in the makatea, making bathing outside the reef a dangerous activity. Low cliffs three to six metres high face the sea everywhere, but there are many recesses in which small sandy beaches can be found. From the top of the sea-cliffs, the land gradually rises to descend again into a swampy area used for the planting of taro and sweet potato. Many dry-land foods are grown in the fertile ground of the slopes between the swamps and the island's top, and include coconuts, breadfruit, bananas, citrus, papaya and mango. The swamp is also the main area where fresh water is found on the island. It is thus not surprising that settlements used to be built here, instead of in the uplands area where the mission erected its foundation, attracting people to the plateau in the late 1900s. The land rises steeply from the inner edge of the swamp to the dome-shaped plateau in the centre, and today people rely on motorbikes to navigate the precipitous incline while laden with produce from the gardens.

Tourism on Atiu, as a result of the rugged terrain, is largely restricted to the adventurous and those interested in geology, a fact that would make the island a remote place, were it not for the frequent comings and goings of its people. Out-migration has been particularly vigorous, with the island losing more than half of its population over a thirty-year period (from a high of 1,455 in 1971). Those left behind are the old and the children, with those who have left visiting and being visited at regular intervals.

Its inhabitants call the island Enua ('placenta') or, fondly, Enuamanu ('land' or 'island' of the birds). The people of Atiu regard their island as the spiritual centre of the Cook Islands, as it was the landing place of Captain James Cook, whose crew went ashore on the island in 1777 to collect fodder for the animals on board. In pre-colonial times, Atiu was known as the core of an expanding regime that, during the reign of Tara-toa, encompassed not just Mauke, Mitiaro and Manuae, but also Rarotonga and Aitutaki. Atiu controlled the system of divine power, by exporting its ariki titles to Mitiaro and Mauke, where the title bearers acted as representatives of the ariki of Atiu, rather than possessing powers in their own right. Atiu's control over the governance of the region remained intact until the early twentieth century, when the introduction of land courts shifted the centre of political control from the tribally held chiefly titles to the system of law. However, by that time a new wave of expansion was already underway, with the establishment of an effective Atiuan community in Tahiti, called Patutoa, where Atiuans had settled to work on plantations at Atimaono and in the capital Papeete. To this day, major decisions such as land settlements or new projects in the Patutoa community require consultation with the tribal leaders of Atiu. The same pattern of political control has arisen in Atiuan communities in Rarotonga and New Zealand, where committees with

representatives from Atiu have been elected to oversee the welfare of the communities and to provide means by which funds can be raised for projects. An example of the success of Atiuan community expansion is the two-storey hostel in Rarotonga, built in 1981 in less than six months, which since has given shelter to Atiuans arriving for the annual Constitution Festival or other community affairs. Today there are many such hostels where Atiuans live, built from money raised by the community.

Atiu is today divided into five villages, five districts and three tribes. The villages flow into one another and are difficult to separate visually, although each tribe tends to be centred in a particular village. Ngati Teakatauira, centred in the Ngatiarua district, is the highest ranking tribe. Its ariki or high chief was the first ariki title created on Atiu in c.1760. The name of the title is Ngamaru Ariki. The tribe of second ranking is called Ngati Paruarangi, its ariki title is Rongomatane Ariki, and its landholdings are centred in Tengatangi. The remaining tribe is Ngati Nurau and its base is in the Mapumai district. The name of its ariki title is Parua Ariki. Each of the remaining two districts is loyal to these three named tribes. The high-title holders in these two districts are called big mataiapo, to set them apart from the high-ranking ariki and the lower-ranking ordinary mataiapo. Again there are named titles for the big mataiapo: Teenui village has the title of Makopi Mataiapo; in the neighbouring Areora village there are three mataiapo titles called Paerangi Mataiapo, Tinokura Mataiapo and Aumai Mataiapo.

The villages and districts in turn are ranked according to their association with the ranked ariki and mataiapo titles. The highest ranking village is Ngatiarua, as its ariki title Ngamaru is of the highest rank; Tengatangi village and district is second in rank, Mapumai is third. Areora village and district is always fourth in rank, because its three big mataiapo are subservient to the highest ranking ariki title, Ngamaru. Neighbouring Teenui occupies the lowest rank. Each of the three tribes today lives scattered across all five villages, with each tribe, however, still recognizing its own ceremonial ground, the marae. The marae is a stone enclosure still used for the ceremonial succession to the ariki and other titles. In the past, important meetings to consider matters of tribal importance were held inside these enclosures. Still, today, only high-ranking people – ariki, mataiapo, rangatira or lesser chiefs, as well as ta'unga or knowledgeable elders – are allowed to enter the marae, although such rules are not applied with the strictness they once were.

Succession to the ariki title is by 'divine choice', though accession is won through achievement and selection from within the immediate family of the predecessor. The title of the mataiapo, however, is divinely ordained and should not be interrupted by death; in a similar way to the medieval kings of France and England, the title of the mataiapo has to be transferred to a successor on the same day the previous title-holder dies. The succession to the title of ariki, however, is left to a meeting of the extended family after the burial, with the decision being passed to the assembled mataiapo of the tribe, who have the final vote. Crucially, the office of ariki is independent of the person holding the title, and survives as representative of the social body, even if no person can be found to take office; the title of the *mataiapo*, on the other hand, requires a human mediator for the transference of the divine office from one generation to another.

With the introduction of land courts in 1905 and the distribution of land to individual people, the partly divine and partly human power of the *ariki* started to decline, while the *mataiapo* took on community related representations. The Island Council, created in 1915, further reduced the influence of the title-holders, as its members were elected representatives, reducing the remit of *ariki* and the *mataiapo* mainly to ceremonial functions. Today, many community affairs are organized and debated at the level of the church. The choice of church allegiance, which would currently be between the Cook Islands Christian Church, the Catholic Church, the Assembly Of God, the Apostolic Church, the Seventh Day Adventist Church and Jehovah's Witnesses, will thus group people into allegiances that are independent of other factors embedded in kinship and land titles. Which church denomination people belong to is in part determined by social allegiances, in that children will commonly follow the religion of their godfathers or godmothers, although it is also common for people to assume a different religion if, for example, a post such as deacon is offered to them in the 'new' church. Some families baptize each one of their many children into a different religion.

These spiritual fields of influence that appear to regulate much of the everyday movement and activities of people on the island of Atiu today, were in the pre-colonial past taken up by image-based polities: Tangaroa, Rongo, Tane and Tutavake. The polities were represented by wooden carvings, which were wrapped in layers of barkcloth and tied with cordage knotted in the shape of flowers to hold much-valued red feathers. The only good description we have of these so called staff-gods is by the early anthropologist Sir Peter Buck (1945). From his description we know the knotted feather-holders were arranged in precise numerical order, seemingly creating a pattern across the surface of the bound sculpture. We do not know more about this pattern composed of flower-shaped feather-holders, but the latter-day floral patterns on patchwork are certainly reminiscent of them.

We also do not know how these staff-gods were used in ceremonies, or what they were assumed to do. The only comparative description we have is from pre-Christian Tahiti, where similar staff-gods were used for a ceremony called *pai' atua* or 'the gathering of the gods' (Babadzan 1993 and 2003). This centred around the unwrapping of the figures (allowing the souls of the dead to be separated from the domain of the living) and their rewrapping (to retain the spiritual source of power on the land). The French anthropologist Alain Babadzan describes the ragged contours of the wrapped staff-god as symbolizing the shroud of the soul travelling in the reverse direction from the dead to the living. We can recall his words, realizing their powerful meaning, when looking at the patched *tivaivai*, which are wrapped around the dead in their graves, and which permanently bind the memory of those who have died to the living.

The practice of 'binding of the dead to the living' is also recalled in a naming tradition known as *ingoa mate*, or the giving of death names, the official registration of which was banned by the Cook Island Christian Church in 1958, but which is still privately observed within families. Names recalling the assumed cause of death, such as Mate Anu ('cold death' – a reminder of death in the cold sea) or Totiko (the name of a passage in the reef in which a young man drowned), are given to family members in remembrance of the deceased.

Teaspoon

One of our first friends on the island was a tiny, elderly man, Punua Tauraa, who lived in our neighbourhood. People called him Papa Tipunu. From him we learned many dos and don'ts. We were grateful for his help in understanding and abiding by local customs. He showed us his family's cave, Rimarau, and told us the spooky legends of cannibals and burials that surround it. We saw him in action at family functions as a great orator; and all women of my generation remember him as a lover of *tivaivai*. Whenever the *vaine tini* (women's group) staged their annual show, he would take advantage of his position as respected elder, and rise to tell us how initially he had resented his wife's begging him for money to buy fabric for new *tivaivai*. In her younger days, Mama Mataua was a keen and gifted *tivaivai* maker, whose works stood out as different, and photos of which even found their way into printed publications. At the end of his speech, he would always admit with a growl in his voice, but a loving and appreciative smile in his eyes, that despite all his resistance to yet another sum of money spent on his wife's whim, the result had been amazing. He therefore encouraged all young women to follow his wife's example and beg money from their husbands to continue the art of making *tivaivai*.

The time came when we started to learn some words in the islands' local language. To my surprise, I found that *tipunu* meant 'teaspoon'. Of course this made me think of Papa Tipunu. So I went to see him, to find out how he had received his name. He told me that it was a death name, given to him in remembrance of his feeding a dying relative. 'But don't you EVER call me "Teaspoon",' he warned me with rolling eyes.

Some days later I needed to bring something to Papa Tipunu. Remembering his warning on the way, I kept reminding myself not to call him Papa Teaspoon. I reached his house. Entirely without intention, when I entered to look for him my subconscious played a trick on me, and I called out 'Papa Teaspoon!' I wanted to dissolve with shame into thin air, or to disappear into a rat hole. But I could neither help my embarrassment at the involuntary mistake, nor contain my laughter at the funny situation. I was much relieved when the old man finally stepped out of his door and joined in with a bright smile, giving me a big hug and waving his hand in mock punishment.

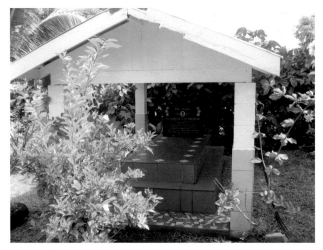
A grave in Mangaia.

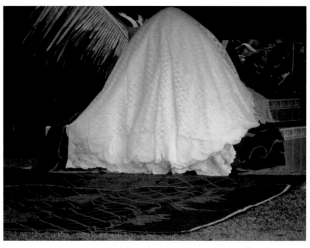
A gravestone covered with layers of *tivaivai* and lace.

Pati-vai or 'beg for water' is one such name, which serves as memento mori, reminding all those who hear the name of the fact that dying is part of life, but also recalling a dead person who once lived as part of the community. The young person who is given the name, and is called for in this way with affection by his family, is made to stand in a definite relation of succession to the dead, as he or she is associated with the

act of offering water, paying back services received during life. In addition to explicit death names, which refer to services performed by the recipient for the deceased and which are acquired later in life, all birth names also refer to a deceased relative or friend of the person giving the name to the child – specifying either the cause of death, some personal attribute of the deceased, or some circumstance associated with the

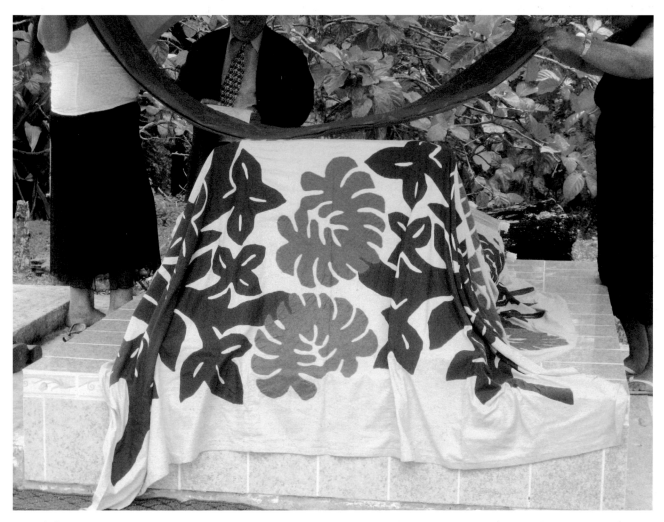
Removal of layers.

last illness of the deceased. Death names can be seen to create the fiction of a social body, which is immortal, as it offers both new physical bodies for the souls of the dead and also cancels out the indebtedness of the living towards the dead.

The dead are also incorporated into the lived-in social space: graves are erected in the form of miniature houses right outside the veranda where people sit and talk during the day. The dead are entombed underground in a cement and stone structure covered by a large rectangular limestone lid, upon which rests a smaller limestone construction, which can be of varying shapes, such as curved, pitched or flat. Occasionally the concrete is embellished with ceramic tiles. At times this stepped monument is topped with a roof built from wood, sometimes made to last and erected in concrete and covered with roofing iron, resembling in shape the house it stands next to. The most important and final addition to the grave is the memorial stone, which is erected a year after the death. The ceremonial uncovering of the headstone is a most important occasion, bringing together family members from near and far. Some twenty *tivaivai*, lace cloths and store-bought blankets will be layered over the tombstone, to be removed one by one by dignitaries and family members invited to the ceremony. This ceremony is in fact a 'second burial', in that the soul of the dead is re-captured and turned into an image which can aid the continuous remembrance of the dead. As graves are always erected on private land, they are to be found in their various stages of construction and decay across the entire island, even where there currently appears to be no sign of habitation.

One would be mistaken, however, to assume the island reflects the tranquillity we may associate with death. In fact, noticing every metre of habitation on the island to be a part of a giant graveyard enables one to realize how effectively death has been turned into a life-giving event. The island, to the surprise of every visitor, buzzes with life and energy. The day starts before dawn on Wednesdays, Fridays and Sunday mornings, with the beating of the big slit gong by the CICC deacons for church service. The *mataiapo*, the town crier, will walk the streets for important calls to meetings, sports activities and the like, any time of the day; though Radio Atiu and announcements in church have started to replace this function. Soon after morning church, elderly *mamas* are racing between house and garden on their motorbikes, carrying babies on their backs, while adults and children alike are engaged in sports activities from dawn till school time and from afternoon till dusk. The year is divided into sporting seasons, from tennis to netball to football, and from running to golf (in Rarotonga). All sport is competitive, with villages competing against each other and in turn against other islands in the Cooks, and participation is taken very seriously. As yet there is no gym on Atiu, but, to judge from Rarotonga, this would be hugely popular, as there the local gyms are packed from 4 a.m. onwards, with the lunchtime and evening classes of aerobics and dance work-outs regularly full to capacity. There is no rest after sun-down, as invariably there is dance or song practice on the island agenda, in preparation for competitions on neighbouring islands, Rarotonga or even New Zealand. The making of costumes for dances, the binding of floral necklaces for those departing or arriving by plane, the sewing of dresses and uniforms for the family to wear at public events or sporting competitions, as well as the busy stitching of *tivaivai* for the many ceremonies attended to by the household and for the competitions organized by the Church, fill the rest of the day.

Genealogy, personhood and the quilt

Where the dead are in close proximity to the living, one would assume genealogies to be short or even non-existent, as the souls of the dead are recycled to the living in every generation. In the Cook Islands this appears not to be the case, for here individual names, even death names, are not passed on to the living. Instead, individual names appear at death to melt into a polity of names whose logical internal structure allows the Many to be conceived as analogous to the One, thought of as the originating ancestor. People can recount with ease as many as 47 generations of their tribe back to the founding grandfather, who is usually remembered by name, and a further 100 generations or so of so-called godly sons.

In order to understand how the remembrance of such extensive genealogical connections is possible, we need to enlist the help of Jukka Siikala's (1991) superb ethnography about myth, history and society in the southern Cook Islands. His data consisted of the extended myths and historical record of the triangle of islands in the southern Cooks called Ngaputoru, which comprise Atiu, Mitiaro and Mauke. Situating himself on Mauke, Siikala describes how mythology captures the hierarchical relationship between the island of Mauke and the island of Atiu, in that Mauke is recounted in myth to have given wives to Atiu in exchange for access to divine power or *mana*, vested in the *ariki* titles of Atiu. The hierarchical relationship was maintained over time and kept in balance as a result of the relationship between these two islands and the third, called Mitiaro, where in mythology the female children of Mauke women are moved in adoption (as tribute for support in warfare) to grow their own offspring, who return again in marriage to Mauke. The relation between the three islands is thus an asymmetrical and yet transitive one, in that each island is enchained with another through the movement of women in exchange for the chiefly titles, which both Mauke and Mitiaro are granted by Atiu.

In his detailed examination of myth, Siikala isolates the logical structure comprising the cosmogony of Ngaputoru, at the centre of which is, surprisingly, not chiefly power, but alliances created by marriage ties reckoned through women. The reason for the importance of what Siikala calls the 'female element' is its ability to create an autonomous, 'new' and extended social structure, which over time has come to include not just Ngapatoru but 'the whole British Empire'. Factors in this process include the size of the islands, which require people to move out in every generation, and the 'double character' of the divine origin of an *ariki*, who comprises two concepts of power in one office, one divinely produced, the other humanly produced (Siikala 1991: 132). As in ancient Hawaii, whose system of social organization was described by the anthropologist Valerio Valeri (1985), in Ngaputoru there are two prerequisites of chiefly succession: descent and effective alliance. According to Valeri's interpretation, effective alliance secured through marriage is more important to succession than descent, because 'the supporters of the predecessor are likely to be different from the supporters of the successor, since the latter, to have any chance to succeed at all, must put together a faction long before the problem of succession arises'. In this building of factions, the difference between patrilateral reckoning and a matrilateral one is very marked in Hawaii. In Mauke, according to Siikala, the small size of the population led to descent being reckoned at once through an older and a younger brother, creating an optative system in which chiefly succession and the continuity of the tribe depend upon the mediating role of women. The *ngati* or tribe is thus structured by interlocking fields of relations created through the movement of women in marriage between the islands.

Myths, which recount the history of a tribe, are shown by Siikala to utilize the same hierarchical structure, at the base of which lies the incompatible relation between divine authority and human authority combined in the office of chiefly power. If we accept that Valeri and Siikala have correctly interpreted the cosmogony of the Cook Islands and Hawaiian kinship systems and their tendency for asymmetrical hierarchical structures that replicate themselves logically over time, we can see how this would enable the recollection of extended genealogies.

We could assume that the introduction of Christianity changed the triangular and hierarchical structure of cosmogony and society forever. Indeed Siikala argues that the spatial segregation of people inside the church on the basis of gender, with men and women facing each other as equals when attending church, nearly provoked the breakdown of the traditional exchange system and its concomitant social order, which depended upon the unequal relation between wife-givers and the givers of divine power.

Siikala does not explain what happened in Cook Islands society so that exchange-based relations continued to inform the flow of power and of wealth, which to this day connects islands and islanders across diverse nations in most effective ways. He merely speculates that wife-givers had an advantage in maintaining and building upon relations of exchange, in that their offspring came to be dispersed not just across Ngapatoru, but also across

the 'British Empire'. However, by not mentioning, let alone examining women's production of *tivaivai* for exchange, he arguably missed an important key to understanding the relation between conversion and the reorganization of society. It is argued here that both the material and the social history of the *tivaivai* bring to light the transposition of the social nexus, which consisted of the exchange of divine power for human aspiration, onto a material nexus, composed of the harnessing of uncut, imported cloth and the imposition of creative agency which transforms this cloth into memorable patterns that can be given away to underwrite relations of allegiance and friendship as much as relations of kinship.

The importance of uncut, unpatterned cloth across the Cook Islands can be explained by its analogy with divine power, which, when harnessed, is substituted for and translated into the generative agency of women, who as mothers assume the task of transforming this cloth into numerous patterns through the stitching of *tivaivai*. Like the mother who in marriage left the island to bring up her family elsewhere, *tivaivai* are rarely made to stay within the household, but move, at least temporarily, elsewhere, usually following children and grandchildren. Because the thus harnessed cloth is still in its cut stage associated with divine agency, bearing a likeness to the duality of being which characterizes the *ariki*, *tivaivai* are rarely acquired as commodities. Instead they are inalienable gifts, to be returned on the death of their prior owner, remaining with the tribe in just the same way as their *ariki* titles do. Having said this, it is important to note that some *tivaivai* are bought, and that some others do not return to their previous owner, but are given to another family member. Women, for example, tend to give *tivaivai taorei* to their daughters as wedding gifts, destined to be wrapped around their husbands on death.

Each lock of hair removed from a boy (made to sit on a *tivaivai*) at his hair-cutting ceremony represents the divine ancestry of his father's tribe. As it is cut and paid for by his

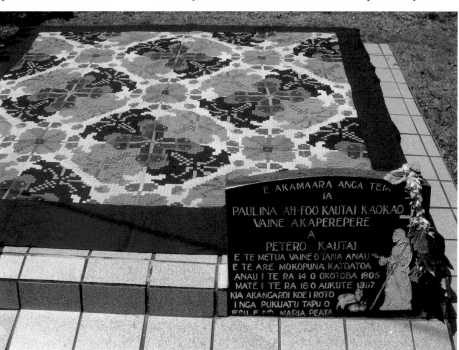

A *tivaivai taorei* designed and sewn by Paulina Kautai[†] and her sewing group in 1960 as a wedding present to her daughter-in-law Ngakura Kautai, placed on Mama Paulina's grave.

Pakotianga rouru: the hair-cutting ceremony

The first important event in a Cook Islands boy's life is his hair-cutting ceremony. This is an event that was introduced by European missionaries, whose apparent intention was to improve the natives' hygiene and appearance; at least outwardly (Buck 1927: 104; Sunderland and Buzacott 1866: 238; Maretu 1983: 69).

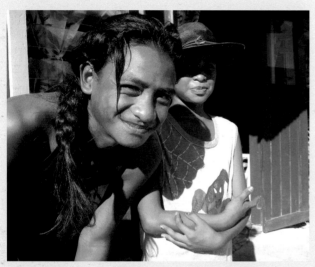

A teenage Atiu boy with braided hair awaiting his hair-cutting ceremony, and his younger sister.

One of our New Zealand Maori hosts at a *marae* in Rotorua told me many years ago that one must never sit on a cushion, as it was indecent to desecrate the place to rest one's head with your behind. We were also told in New Zealand that some people, even today, believe that body parts such as nail clippings and hair must not be thrown away carelessly, but rather buried in a secret place, lest an evil-spirited person might use them for sorcery. The pre-missionary Tahitians held similar beliefs with regards to hair, nail clippings, saliva and food leavings in which, they claimed, some of the person's *iho* (essence or nature) lived and which a sorcerer could use to lead his spirit accomplice to the victim to do him harm (Oliver 2002: 46).

Considering the importance that Polynesians gave to their heads and hair (Maretu 1983: 23; Oliver 2002: 111), it was to ask a great sacrifice of the men to have them openly document their conversion to the new religion by cutting off what, in their view, may have constituted their masculine beauty. Giving away their hair clippings was perhaps meant to demonstrate their belief in the new religion and their denunciation of the superstitions of heathen sorcery.

The hair-cutting ceremony can be celebrated at any age, but it is reserved for boys, and is generally an honour given only to the first-born. Once the family has decided to grow the boy's hair they will start preparations. These may include establishing special food plantations, and raising goats and pigs, so as to feed the guests; sewing shirts and dresses for the family members and the special boy to wear; and, of course, sewing *tivaivai*. Depending on the social status of the celebrating family – or the importance this family wishes to demonstrate – the celebration can range from a small ceremony at home in the company of the immediate family and friends to an extensive feast with hundreds of invited guests held in Atiu's main hall. Often the age at which the boy undergoes the ceremony is advanced, enabling more and longer hair to be grown for cutting, and allowing the family longer to prepare for the big event. The special boy's hair is left entirely uncut before the event. For reasons of hygiene and comfort it is worn in a plaited braid at the back of the head, neatly tied with a string or ribbon.

While sometimes the boy's features clearly indicate his gender, despite the hair style, sometimes it is not so easy to tell. I remember my confusion when I asked the little girl of our friends for the name of her sister. 'This is not my sister, he's my brother with long hair!' was the indignant answer.

As in all important celebrations, *tivaivai* decorate the walls of the room or hall. For the special boy, an arm chair, small sofa or fancy chair is covered with a beautifully finished *tivaivai*. Sometimes, *tivaivai* even cover the floor over which the boy thus honoured has to walk to his chair, so that his feet do not touch the ground. The boy's hair is tied with ribbons into as many strands as there are important people given the privilege to cut a strand by the family.

While prior to this day the boy would be 'allowed' to behave in girlish ways, this is the moment when he 'becomes a man'. It can be amusing to see how hard some of them try not to cry,

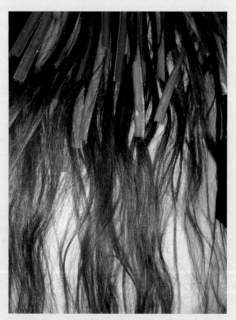

The boy's hair is tied into as many strands as there are visitors invited.

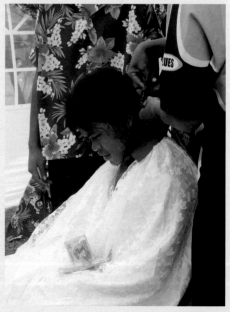

The cut-off strand of hair is kept for memory's sake.

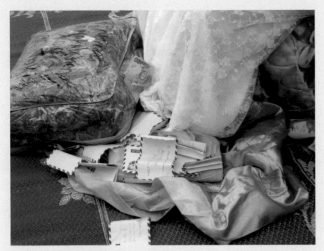
They leave a present at the boy's feet.

the stubbly hair into a decent shape. Greeted by the applause and cheers of the crowd, the cloak or cape worn during the hair-cutting is taken off, and the young man stands up. A family member, or the boy himself, still covered with flower garlands to which money has been stuck, says a few words of thank you. Grace is said. Now the visitors are invited to enjoy the rich food that loads the tables in front of them. Sometimes it is hard to recognize the young man on the street the next day, proudly sporting his fresh haircut with a new macho air about him!

when over a hundred people pull at the strands of their hair to cut them close to the scalp, more so when the boy is very young and the hair has not yet grown so long. After a short prayer and some introductory words, the names of the privileged are read out one by one. They come forward and perform their duty, keeping the strand of hair and leaving, in turn, a present at the boy's feet. During the ceremony, the boy's genealogy will be related to the visitors. Sometimes family members dance up to him with presents of mats and *tivaivai*, sometimes almost burying him under textiles. Once all the strands have been cut and all the presents have been brought forward, the father, or a family member who is versed in the art of hair-styling, will cut

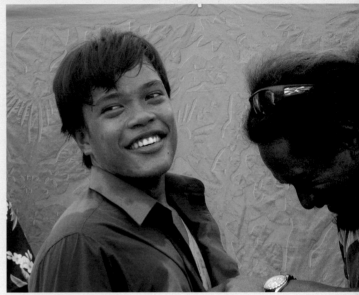
Relieved, the boy stands up as a man.

mother's family, he is told the genealogy of his father's tribe. While he listens to the history of his tribe, he is made to touch the embroidered threads and patches of cloth on which he sits and which are partly wrapped around his body. Now a 'new' man, the transaction of the lock of hair brings relations of mutual responsibility with his mother's relatives. As a symbol of the divine power of the tribe, in its uncut stage boys' hair ties them to a phase of inactivity in which they do not have to take part in exchanges with maternal kin, often for more than a decade and sometimes into early adulthood. The lock of hair, which transfigures into money through the act of exchange, impresses on young men the responsibility brought by a duality of being in which a person is regarded as possessing innate power that becomes socially active only as it is externalized in material form.

Women, by contrast, have no first haircut to be remembered. In fact, in pre-contact times, Cook Island women wore their hair short, rather than long, as the stereotype of the 'dusky' maiden in tourist photographs would have it. The reason, quite simply, is that they are not burdened by the contradiction born of inherited and acquired dispositions, but can acquire new social relations through their own actions. The life-stories recounted by women in the book put together by Lynnsay Rongokea (2001) contain many tales of girls

adopted by their grandparents, and learning, like Vereara Maeva, to sew and weave baskets, to sing and dance and grow themselves into *ta'unga* or knowledgeable elders. Such *ta'unga* have the clear advantage of 'the female element', as Jukka Siikala put it, in that they can harness, transform and grow social relations throughout their lives. As they purchase cotton cloth and thread with money, they harness its power and make it their own, no different in principle to the symbolically female island of Mauke which a long time ago acquired the titles of *ariki* in exchange for affinal relations.

Patchwork cloth, made of hair-like threads, like the coarsely woven Azlin which is preferred by Cook Islands women, and the multicoloured cotton threads which are popular for sewing and embroidery, clearly succeeded in reorganizing Cook Islands social structure in post-contact times. It did so, one can argue, not just because of prior existing conceptions surrounding fibre, as many anthropologists have argued, but mostly because of its material and visual analogy with the cosmogonic constitution of personhood and society.

The following biographies of Cook Islands women relate not just how they learned to make *tivaivai*, but how they conduct their lives within the busy intersections that are created by ever expanding fields of relations. Whereas men tend to stay, either at home or abroad, like their *tivaivai*, women move back and forth, and further afield, quietly and patiently, and make the society that came to form the basis of the Cook Islands nation.

'A Quiet Life' by Vaine Wichman

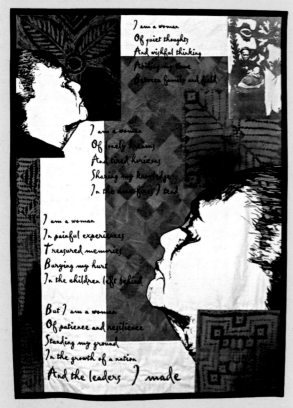

E vaine au ma te au manako muteki e tetai au manako umumumu, ma te tatuatau I toku taime I rotopu I took ngutuare tangata e pera taku tanu kai.

E vaine au ma toku au manako moemoea e te roriroi tikai, te tua nei au I toku karape na roto I te auau o tako umu.

E vaine au I roto au tuatau mamae kua akamaara katoa au I toku manuia. Uuna atu I toku mamae kit e tamariki e ora nei.

E vaine ra au ma te akakoromakianga roa e te ngaueue kore. Te tu tango nei au I roto I te akatupuanga o te basileia e tona au arataki k otaku teia il akatupu.

I am a woman of quiet thoughts and widespread thinking, dividing my time between family and field.

I am a woman of lonely dreams and tired horizons, sharing my knowledge in the *umu* fires I tend.

I am a woman of painful experiences, treasured memories, rearing my hurt in children left behind.

I am a woman of patience and perseverance, standing my ground in the growth of a nation and the leaders I made.

Above: art quilt by Andrea Eimke, now in a private collection.

Tivaivai taorei designed and sewn by Akaiti Ama[†], now in the collection of the Rarotonga Resort Hotel.

Women, their biography and their *tivaivai*

Akaiti Ama was the eldest of five children in an important noble family on Rarotonga. On her mother's side she descended from Potiki Taua, a high priest appointed by Tangiia Nui Ariki to perform priestly functions, especially concerning the investiture of *ariki* (high chiefs) at the *koutu ariki* (chiefly court) named Arai-te-tonga-vare-roa-paa (see Savage 1980: 266). According to the genealogical records studied by Stephen Savage between 1894 and his death in 1941, and posthumously published in 1962 in his now classic *Dictionary of the Maori Language of Rarotonga*, this high priest also held the title of Teariki Taraare Mataiapo, one of the seven *mataiapo* titles of the district divisions which today make up the district of Avarua. On her father's side, she descended from an important chiefly line from Tahiti, Tamarua Nui.

Akaiti Ama gave birth to nine children, but also taught throughout her married life: first as a schoolteacher and then as a high-school principal. In her later years she performed the important function of Justice of the Peace.

Alongside her busy public life, she managed to design and sew *tivaivai* for her children. When she was pregnant with her last-born son, Akaiti kept her husband company in the family's pineapple plantation, an experience that inspired her to design and sew an unusual *tivaivai taorei*, which has survived her death. When she suddenly died in November 2003, her grave was erected beside the ancient *marae* of Arai-Te-Tonga, which lies beneath her son's house. Her younger sister, Terangi Little, has taken over as head of the family and can recount 47 generations of ancestors.

(continues page 35)

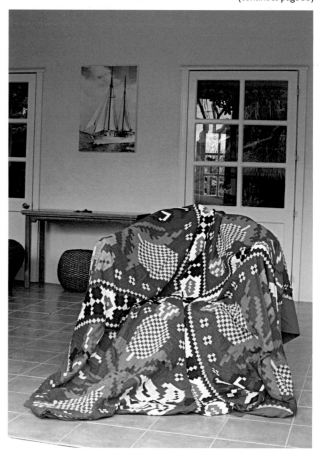

Tracing *ta'unga*

When I first became familiar with the art of *tivaivai*, I was quite surprised to discover that most women did not design their own work. I learned that each of Atiu's five villages had one or two *ta'unga* who would cut the *tivaivai* for the other women. You took the fabric to the *ta'unga* of your choice, mostly the one that lived in your village, and decided upon a motif that you would like them to cut for you. Once cut, you and your friends, if necessary, did all the work, or you could pay a skilled seamstress or embroiderer to finish the *tivaivai* for you. If the *tivaivai* were displayed, for instance in an annual show, only the owner's name mattered, and the praise, if any, was given to her. Naturally, in such a small community as our island, everyone knew who had designed, cut and sewn your *tivaivai*! Over the years I learned to distinguish the individual styles of the various *ta'unga* on Atiu.

I think it unlikely that the art of making *tivaivai* came to Atiu before the end of the nineteenth century, and if *tivaivai* themselves did, they would have been as a gift to the *ariki* (high chief). *Tivaivai* making was probably not common amongst ordinary women until the beginning of the twentieth century. My first informants in the early 1980s were elderly Atiuan women, most of them no longer alive. They all attributed the introduction of *tivaivai* making in organized groups to Teakarua Moni, though some admitted that there must have been some other women who already knew how to make *tivaivai*. There is nobody left to ask for confirmation, nor could I find any written records.

Teakarua Moni

Teakarua was born in Tahiti in approximately 1900 and died in New Zealand around 1982, the exact dates being unknown to her nieces on Atiu.

Teakarua came to Atiu with her parents. Her father was said to have been a Chinese merchant. She most probably learned the art of making *tivaivai* from her Tahitian mother. As a young woman she went back to Tahiti, bought all

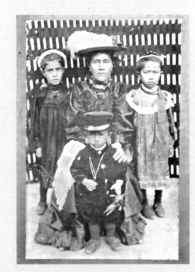

Teakarua Moni (left), shown here with her mother, sister and brother, John.

sorts of merchandise, and shipped it to Atiu, where she and her brother John set up the first privately owned retail business on the island. After her return from this trip, she started gathering young women around her and teaching them all sorts of craft skills. Thus, the first women's sewing group was formed. The *tivaivai* sewn in these groups were *taorei* (mosaic piecework) and *manu* (snowflake-style cut-out appliqué). Teakarua Moni's niece still keeps as a family treasure a notebook in which her aunt recorded the patterns for several *tivaivai*, the oldest entry dating as far back as 1936. She also owns the *tivaivai* to go with that pattern, keeping it well protected in a large metal storage chest.

Emily Teariki Rakei (Mama Puni)

It is the end of this year's Christmas holidays. Atiuan visitors from overseas are leaving the island again to return to their New Zealand and Australian homes. I will have to rush if I want to catch up with Mama Puni, who now lives in Wellington with her children while her daughter Teau looks after the family home on Atiu. Many of the family heirlooms are still stored here in large metal boxes secured with big padlocks.

Her smile is still as friendly as I remember it from last year's visit, her voice still sounds like the tomboy she must have been in her younger days. 'What do you want?' she teases me. I have visited her so often, borrowing *tivaivai* for overseas exhibitions, or asking for information, like now.

Mama Puni was born on Atiu on 24 April 1936. She is proud that she will soon turn seventy (this interview was conducted in 2006) and feels as fit as a young woman, she assures me. She looks it, too! As niece of the founder of the first sewing group on Atiu, she learnt her skills of designing and cutting *tivaivai* from her mother and her famous aunt Teakarua Moni. Under her mother and aunt's supervision, she cut her first *tivaivai* when she was twenty-two, and has cut over a hundred *tivaivai* since for the women on Atiu, mainly for members of her Teenui village sewing group and her own private requirements.

I ask her about the trophies I can see proudly displayed on a side table. She picks one up and tells me that she won it for her participation in a craft exhibition on Rarotonga in 1992. In 1995 she was elected president of the Teenui women's group. 1995 saw a memorable *tivaivai* exhibition on Atiu's Boys Brigade Ground. Two pastors' wives were leaving the island at the end of their

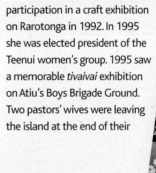

Right: Emily Teariki Rakei, alias Mama Puni, proudly posing in front of Teakarua Moni's *tivaivai* in 1994 and displaying her family photographs.

Below: Precious *tivaivai* are stored in securely lockable metal boxes.

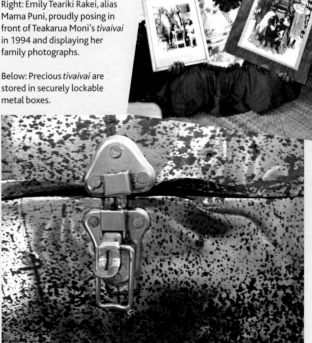

husbands' term in office, and they designed the programme for the exhibition. In addition to an embroidered *tivaivai* and a matching pair of cushion covers, the members of the Atiu *vaine tini* (women's group) also had to make a matching sofa cover, two cushions, a painted pandanus mat and a door mat! Mama Puni was kept busy that year, and earned her position as a well-respected *ta'unga* for her village.

Several of her works have been exhibited overseas, and a *tivaivai* of hers is now to be found in the collection of the British Museum in London. You can see it on the cover of this book. Her greatest treasures are two *taorei tivaivai* that date back to approximately 1936 and were designed by Teakarua Moni.

Mama Puni now lives most of the year with her children in Wellington, coming home to visit Atiu only for Christmas holidays. Even at her advanced age, she still cuts the occasional *tivaivai* for neighbours and friends, and continues sewing her creations for children and grandchildren. I hope to see her again next year...

Tuatai Koronui

I love Mama Tuatai's work for its unique and creative use of uncommon motifs. Every *tivaivai* has a particular personal touch, with swans, crabs and birds populating her works. She has a vast knowledge of different embroidery stitches and enjoys creating her own. Her fantasies are witty: some of the flowers on her *tivaivai* show insects sitting on their petals.

Tuatai Koronui was born in Tukau on the Northern Group island of Manihiki on 9 November 1938. Already interested from an early age, Mama Tuatai learnt her skills from her mother, Tina Morara, and her relative, Mrs Ngaupoko Ripata, when she was fourteen. It is no surprise to me that she fell for the charms of Papa Vaine Moeroa, who is a well respected *ta'unga* in many areas himself, and who has been a rich source of knowledge for his students at Atiu College during his years as a mathematics and woodwork teacher, and to organizations such as the Natural Heritage Project of the Cook Islands. He won Mama Tuatai's heart, they married and the couple came to live on Papa Vam's home island. In 1964 she cut the first *tivaivai* pattern of her own design, which she still owns. Since then she has cut countless *tivaivai* on Atiu and Rarotonga, many for her village of Mapumai and the members of her Seventh Day Adventist church congregation. She is also an expert in other traditional crafts skills.

Several of her *tivaivai* have been included in exhibitions overseas. We can also see Mama Tuatai in a BBC documentary, *A Little Film About Tivaevae* (1995). She shows us a bedroom in her house decorated with a beautiful *tivaivai* on her four-poster bed, with a matching canopy. It is my personal favourite and belongs today to her son Charles, who was Atiu's Island Secretary for several years before resuming overseas studies to expand his professional career.

Recently her and her husband's failing health made it necessary for the couple to move to Australia, where they now live with their children. I miss her creative touch in our island's annual *tivaivai* exhibitions!

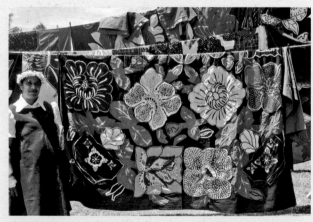

Above: Tuatai Koronui, shown here in the early 1980s with one of her creations, a richly embroidered *tivaivai tataura*.

Below: A detail of one of Mama Tuatai's *tivaivai* showing insects on the flower petals.

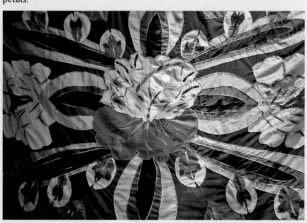

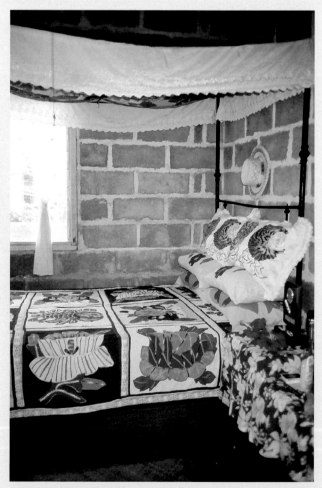

Above: Mama Tuatai's special *tivaivai* and matching canopy.

Patikura Jim

Mama Patikura was born into a family of *ta'unga* on Atiu on 12 June 1948. Her Atiuan father was a carpenter, who had been in charge of the carving of beams and posts for the Cook Islands Christian Church building and the former Areora meeting house. As is common for a first-born child, she was raised on Atiu by her paternal grandmother, Terii Ruia Matunga. Mama Terii was a seamstress. From an early age she involved young Patikura in helping her sew garments. The leftovers of colourful shirt and dress fabrics, mostly cotton printed with floral designs, were made into patchwork *tivaivai* (*tivaivai taorei*). This was Patikura's first exposure to patchwork.

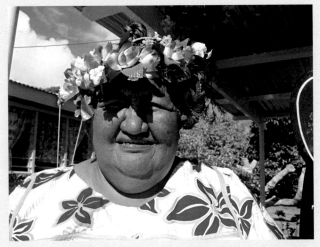

Mama Patikura Jim.

In those days knowledge was even more closely guarded than it is today. Mama Terii was a good seamstress, but she did not know how to design and cut snowflake-style *tivaivai* (*tivaivai manu*). She soon realized her granddaughter's talent, and made her find out about the secrets of pattern design. This was a hard task for a young girl, as nobody wanted to share their knowledge with her. 'It wasn't like today,' she says, 'each village had their own *ta'unga*, who would keep her patterns and skills secret from everybody.' So she decided to learn by tracing around leaves, taking apart flowers and studying her environment carefully, in order to discover the secrets of design for herself. 'Between fourteen and twenty I just practised,' she admitted, 'and the patterns weren't really that good until I was about twenty years old.'

When she was sixteen, she was sent to Tereora College on Rarotonga. It was during those years that her birth mother, Iomai Fariu, taught her embroidery. 'Only a few stitches at a time,' she grinned, 'because my mind wasn't really on the embroidery that time.' We had a good laugh, because on Atiu that means that she had boys on her mind instead!

Her family was not very affluent and there was no money for material to waste in experiments. Patikura remembers having used the children's nappies and some left-over printed material for her first trials in pattern-making. But by around twenty years of age, she felt that her designs were good enough to make decent *tivaivai*.

She did not start designing and cutting *tivaivai* for others until the early 1980s. I met Patikura when I started living on Atiu in 1983, and the first *tivaivai* I ever saw on Atiu was one of her creations: a *tivaivai manu* with an intricate lime-green *kape* (giant taro, *Colocasia macrohiza*) pattern on a royal-blue background that was spread out ready for tacking on the floor of her large family home.

Her beautiful patterns soon found admirers. In 1992, the Cook Islands government employed her as Assistant Women's Officer. She received payment for cutting *tivaivai* and making crafts for and with Atiu women, and even travelled to other islands to cut *tivaivai* for the women there. Her position was abolished in 1996, when the government downsized the public service.

That did not mean Patikura stopped cutting *tivaivai* for other women. Her talent was so popular and her patterns so beautiful that her skills remained in high demand. As many of the older *ta'unga* have died or left Atiu for health or other reasons, she is today the most prolific (and almost only) representative of her profession on the island. That has kept her very busy. She told me that in 2006 she cut close to 100 *tivaivai*.

I asked her if she had passed on her knowledge to her daughters. Pauline, who is twenty-six and lives on Rarotonga, knows how to sew *tivaivai*; and twenty-two year old Joy, who lives on Atiu, is a promising talent who has learned how to cut and is starting to develop her own patterns with her mother's help. Patikura has started to draw the main motifs on plastic floor-covering for Joy's use, to help her develop her own *tivaivai* patterns when her mother is not on the island, thus continuing the family trade.

Patikura has designed a plethora of beautiful patterns using motifs such as orchids, roses, heliconia, lilies, hibiscus, grapes,

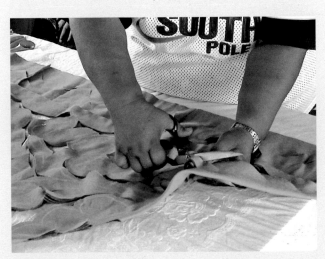

Patikura Jim cutting a *tivaivai manu*.

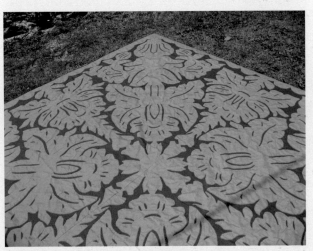

Tivaivai manu (detail) designed and cut by Patikura Jim.

passion-fruit flowers, fans, crowns, butterflies, pansies, chandeliers and candles, and even one showing a woman, which she made for her daughter Ngaua's twenty-first birthday. In 1992 she was commissioned by the ministry of culture to design two 2 × 4 metre *tivaivai* to be used as decoration for the newly inaugurated auditorium during the South Pacific Art Festival. She prefers to design *tivaivai manu*, because she enjoys the challenge of creating intricate patterns. But her colourful *tivaivai tataura* designs are also very popular with Atiuan women.

Patikura is also skilled and knowledgeable in other traditional crafts, such as hat weaving and making costumes from tapa cloth. Many of the tapa costumes worn on Atiu at official occasions, such as openings and investitures, are her work. A number of her creations have been shown in international exhibitions around the world; she was invited to demonstrate her craft at the Auckland City Art Gallery's major *tivaivai* exhibition in 1995 and she also featured in the BBC documentary.

Patikura is aware of her special talent and resents the copying of her patterns. It saddens her that often those who have commissioned her to cut a *tivaivai* for them forget to mention her as the author of the work. 'Without my talent,' she observes, 'those *tivaivai* would not exist! People think that designing and cutting a pattern is an easy job, but it is not. It is hard work.'

Nga Mokoroa

Auntie Nga, as everyone calls her on Atiu, was born on Rarotonga on the 11 March 1936. 'Our houses were next to one another,' she told me, 'there was ours, then there was my auntie's, my dad's cousin's, all on the same family land in Nikao.' It was this cousin of her dad's, Mrs Moeava Tukutamaki, who took her along to her sewing group when young Nga was only fifteen years old. 'We were a quite well-off family, because my father was employed by the Education Department as a driver. So there was enough money available to afford fabrics to make *tivaivai*.' The older ladies needed help with a patchwork *tivaivai* (*tivaivai taorei*), work which is usually shared by a group of eight women. They had only two weeks to finish it, very little considering the huge amount of work required to join thousands of one-inch squares. Nga was quick to understand what was required and proud to be asked to help.

'The *ta'unga* would draw a motif and it would then be divided into squares which could be counted. That way she would be able to create a pattern.' Nga watched the older ladies and practised her skills alongside them. She learned embroidery from her New Zealand teacher, Mrs Graham, at Avarua School. 'I wasn't very good at doing satin stitch,' she admits, 'but the oyster stitch was my favourite and I also liked the feather stitch': stitches that one can still find exquisitely executed in many of her creations.

I asked Auntie Nga at what age she sewed her first *tivaivai* all on her own. 'Wait,' she said, patting me on the arm, getting up and disappearing inside the house. When she returned she brought a brightly coloured patchwork *tivaivai* of many diamonds joined in concentric squares. 'This is it,' she laughed, 'you can see how old it is. It was easy to make. Seven stitches to each square and then you pull the thread.' With a good laugh she remembers another *tivaivai*:

> Papa Paiere had just finished building our own house. Before we had had no electricity and I had to use a kerosene stove whose sooty flame

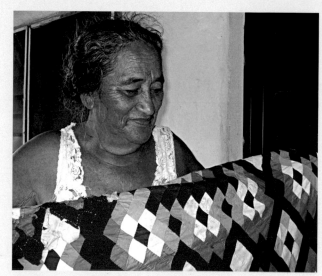

Nga Mokoroa proudly showing her first *tivaivai*.

came up high. I wanted an electric kettle. Then this lady asked me to make her a *tivaivai* with a star pattern; one big star in the centre and eight stars on the outside, four in each corner and four in the middle. I worked hard and got it finished in only two weeks. I wanted the kettle so badly!

When she was twenty years old, just after she had married Papa Paiere, she decided to find friends to form her own sewing group. 'The others were working,' she explains, 'but I had to stay home, because I was breast-feeding my first born.' Her sewing companions soon discovered that Nga was good at designing patterns. Sometimes the payment for her designs was food, but other times it was fabric. Nga liked this payment the best, and would use it to cut two *tivaivai*: one for the friend who commissioned her and another for herself.

Soon Nga was a respected as a *ta'unga* in her village beyond the small boundaries of her sewing circle. Besides Mama Tata Fariu and Mama Kura Uti, this sewing group also included some European ladies, who worked as educational staff on Rarotonga. 'They were very keen to learn how to make *tivaivai*,' Auntie Nga recalls. When her husband was sent to the island of Mitiaro in 1986 as principal of Mitiaro College, her reputation as an expert seamstress and embroiderer earned her an employment as home economics teacher at the same school. 'They were such skilled girls,' Auntie Nga remembers, 'their sewing was so neat and clean. No hanging threads or knots showing at the back!'

In 1991 the couple was transferred to Papa Paiere's home island of Atiu, where Auntie Nga was employed for some years as sewing teacher at Atiu College and where she also continued her involvement with the Girl Guides, of which she has been a member since 1956. These days, Nga Mokoroa, ex-Girl Guides Commissioner for Atiu and Mitiaro, acts as the Atiu Girl Guides' general advisor and is the co-ordinator for the Duke of Edinburgh Award. For several years now, the two retired teachers have managed their own accommodation facility. Papa Paiere has become a published historian and Auntie Nga continues to cut and sew *tivaivai*, these days mainly for her lodge accommodation and occasionally for visitors who sometimes commission her works. Many of her *tivaivai* have been exhibited in international exhibitions around the world, and one has even become part of the Horniman Museum collection in London.

Vereara Maeva was the eldest child born of a father who was part German and part Rurutuan (French Polynesia) and had ancestral connections to Pukapuka, and of a Mangaian mother with grandparental connections to Aitutaki, Raiatea (French Polynesia) and Tonga. She can recount 13 generations on both sides of her family. Though her mother was born and raised on Mangaia, her grandparents and mother's childless sister lived on Aitutaki. This aunt adopted and raised Vereara as her own child, passing on the grandmother's knowledge of *tivaivai* to the young girl. Vereara became a teacher and married a medical doctor, to whom she gave five children, two of whom today live in Australia. Together, Vereara and her husband took up office first on Mangaia and then on Manihiki. On both islands, Vereara was influential in starting women's sewing groups with her teacher colleagues, forming lasting connections among women which are remembered through *taorei* patterns which women came to share as a result of their joint efforts in making *tivaivai*. While on Manihiki in the late 1970s, Vereara joined a group of four women in sewing a *tivaivai taorei* whose pattern was copied from a *tivaivai* created in 1935 by Ruta Tixier, who had been stationed on Manihiki when she made it. This *tivaivai* later accompanied her son to his grave. The pattern was recreated by Vereara in 2001.

Ruta Tixier neé Maoate was a descendant of a Samoan princess who was brought to Ngatangiia (Rarotonga) by the then Minister of the Cook Islands Christian Church, who was called Tamarua and in whose chiefly household she lived. Ruta was born in 1916. She was raised on Rarotonga, where she married a man who was Tahitian by birth but had been adopted into a Rarotongan family. Upon marriage they returned to stay with his Tahitian family and had five children. When in the early 1950s they returned to Rarotonga, her brother asked her to manage his pearl-diving business on Manihiki, taking her children along. Her eldest daughter Mouparau joined the sewing group together with Vereara Maeva and was active in sewing the *taorei* which was given to Vereara and was inspired by her mother's pattern. Parau still treasures the original *tivaivai* from which the pattern, which consists of 27,500 1-inch squares, was copied.

Parau has ten children, five sons and five daughters. Today two live on Rarotonga, one in New Zealand, two in Australia, three in Tahiti and one in Samoa. Parau was a teacher, but followed her mother to Manihiki for eleven years.

Tokerau Munro neé Dean also has Tahitian connections, through her father, and has Rarotongan links on her mother's side. She has nine children who live today in Australia, Papua New Guinea and Tonga. Like Vereara Maeva, she lives in

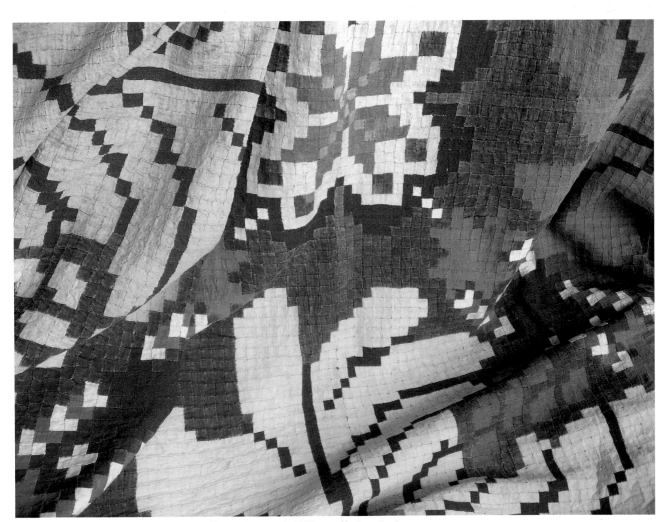

Tivaivai taorei designed and sewn by Ruta Tixier and her sewing group in 1935, owned by Parau Taruia.

Arorangi and has joined a very active sewing group in which she officiates as the designer and cutter. Tokerau is the president of the Girl Guides Association and has travelled widely as part of this office.

One of the youngest active *tivaivai* makers on Mangaia is Patricia Makike Glover. Both her parents are Mangaian, but her paternal grandfather came from Switzerland. Together with her Mangaian husband, she lived in Melbourne (Australia) from 1998 to 2005. She returned with her three children to her husband's village, where she has lived until today. On both sides of her family she is closely connected to the two most important *tivaivai* makers on Mangaia today, who have taken her into their sewing group, teaching her the skills of embroidery and crocheting, and the cutting and sewing of *tivaivai*.

Krudean Kareroa is Mangaian, with a part Aitutakian part Mangaian mother; her father was from Penrhyn and was half Rarotongan and half New Zealander. Her own husband is from Mangaia and has a Rarotongan mother. In addition to her nine birth-children, she raises a number of adopted children.

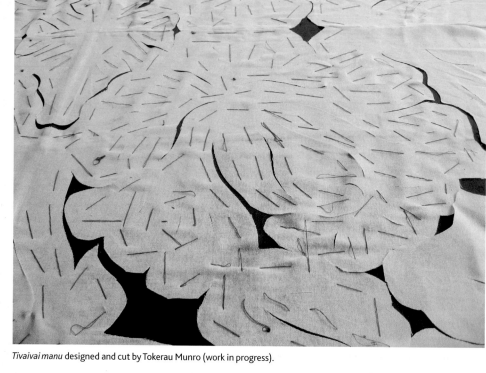

Tivaivai manu designed and cut by Tokerau Munro (work in progress).

She is also taking responsibility for a large number of so-called 'feeding children', whom she raises on behalf of her brothers and sons. Krudean is famous throughout the Cook Islands for her intricate embroidery inspired by children's books. She took up sewing when obliged by her mother to settle down from her beloved work in the garden. She innovated many stitches and collected these in a booklet. She has also designed and cut numerous *tivaivai* for her own family and upon request from other people.

Krudean's friend and relative, Kimiora Samuela, is the oldest of nine children. Her Atiuan father was adopted by an Atiuan woman who lived on Mangaia. Her Mangaian mother was an outstanding crafts-person, who used to sew clothes for her family and other people. Mrs Sam, as she is known, started sewing in 1963, when she returned from Rarotonga, where she had trained as a teacher together with Vereara Maeva. In 1964, she started working on *tivaivai taorei* with four other ladies. Vereara joined her sewing group during the 1970s, when she came to stay on Mangaia. She is the secretary to

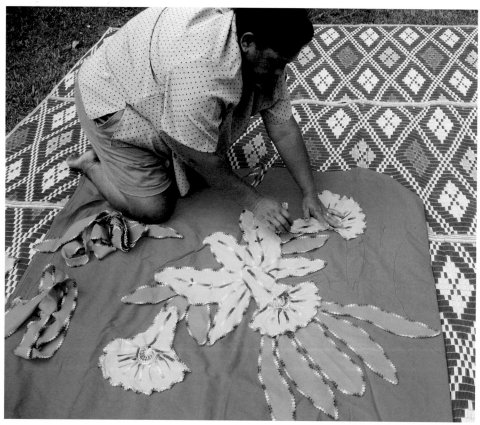

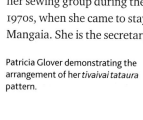

Patricia Glover demonstrating the arrangement of her *tivaivai tataura* pattern.

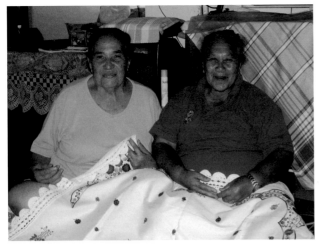
Mrs Krudean Kareroa (left) and Mrs Kimiora Samuela (right), with a tablecloth embroidered by Mrs Kareroa.

her *vaine tini* (women's group), of which there is one in each of the three villages on Mangaia. She is also the secretary and Mangaian representative for any inter-tribal affairs relating to Ngaputoru (the common name for Atiu, Mauke and Mitiaro). Mrs Sam is famous across the Cook Islands for her never-repeated *tivaivai* pattern innovations, for which she draws inspiration from the natural environment of her home island.

Patikura Jim is Atiu's expert in the designing and cutting of *tivaivai*. Born of a Rarotongan mother who had married a Chinese trader, Patikura was raised in Atiu in her grandparents' family as the first-born grandchild of the first-born son. Although her father's father was a true Atiuan, her father's mother was raised in Tahiti, where she worked as a tailor and seamstress. Patikura learned her cutting skills from her grandmother, and remembers having seen *tivaivai* cut from dress material during her early years. Both her father's and her father's mother's family line descend from important *ta'unga* family lines. Her father was an important carver on Atiu. On her father's mother's side she descends from the line of Ngati Paerangi. The paternal grandmother was a renowned *ta'unga* of patchwork. Patikura has cut innumerable *tivaivai*, in one memorable year more than 300, both for her family and for other women in Atiu and other islands. Her husband, whose gravestone unveiling ceremony took place in 2007, worked in Sydney for many years, where he lived with five of their seven children. Today, she works as a women's officer and raises the two children of her daughter who works in Rarotonga, together with her youngest still unmarried daughter.

Nga Mokoroa lives with her husband on Atiu. She was born in Rarotonga of a Rarotongan father and an Aitutakian

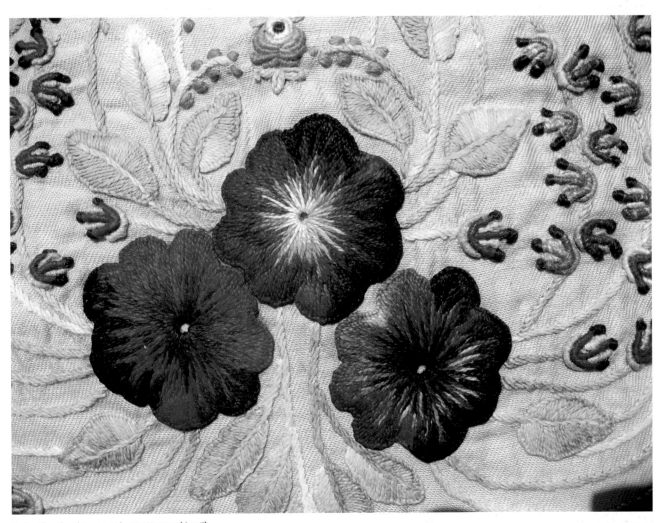
Embroidered cushion cover by Mrs Kareroa (detail).

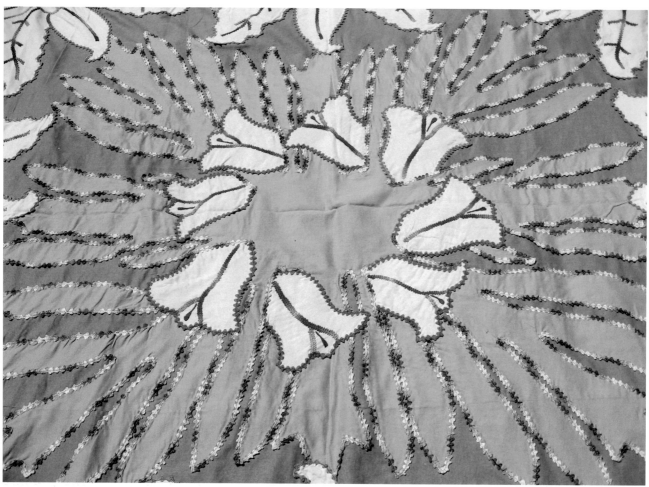

Above: *tivaivai tataura* designed and sewn by Mrs Kimiora Samuela (detail).

Below: Patikura Jim at her husband's grave, displaying one of her hand-embroidered *tivaivai*.

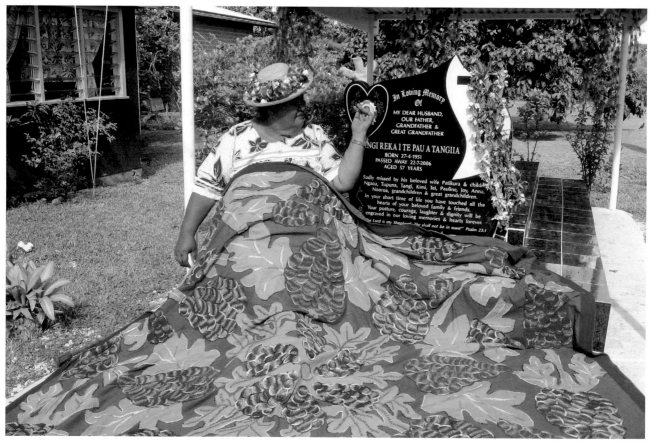

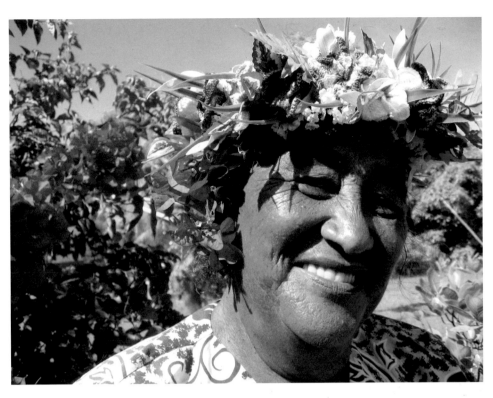

Aunty Nga Mokoroa.

mother. Her paternal grandmother came from Mangaia, while her maternal grandfather came from French Polynesia and the grandmother from Atiu, although her own mother was raised by the mother's grandparents on Rarotonga. Nga had two sisters and six brothers, of whom three still live in New Zealand. The children of her brothers and sisters all live in New Zealand, with the exception of three children of one brother who live on Rarotonga. Nga herself has nine children, three girls and six boys, of whom four live on Rarotonga, two in Australia and two in New Zealand. With two exceptions, her grandchildren have married back into Cook Islands society, and live in the islands. Together with her husband who was a school principal, she taught and lived on Mitiaro and Rarotonga, returning to Atiu upon their retirement from teaching to start an accommodation business. Nga has made numerous *tivaivai* for her family. She is a skilled and enthusiastic embroiderer and she also is renowned for designing innovative patterns.

All these women are of roughly the same generation, one in which *tivaivai* has played an important social role. As keepers of knowledge, they continue to play a crucial role not only within their own families, but also within the wider context of Cook Islands society. Their biographies show us how *tivaivai* form a common thread connecting women to each other and to an unfolding future. For it is notable that it is always with reference to possible futures, to yet-to-be-realized patchworks and gifting relations, that women stitch *tivaivai* into the conversation about their lives. While one might assume recollection to be the tracing of pasts already lived, the memories that are attached to *tivaivai* are, significantly, only mentioned if they pertain to the paving of yet-to-be-lived lives in meaningful ways. It is this present- and future-orientated intentionality of *tivaivai* that has, uniquely, given it the capacity to be at once a 'social fabric' and an articulation of 'modernity'.

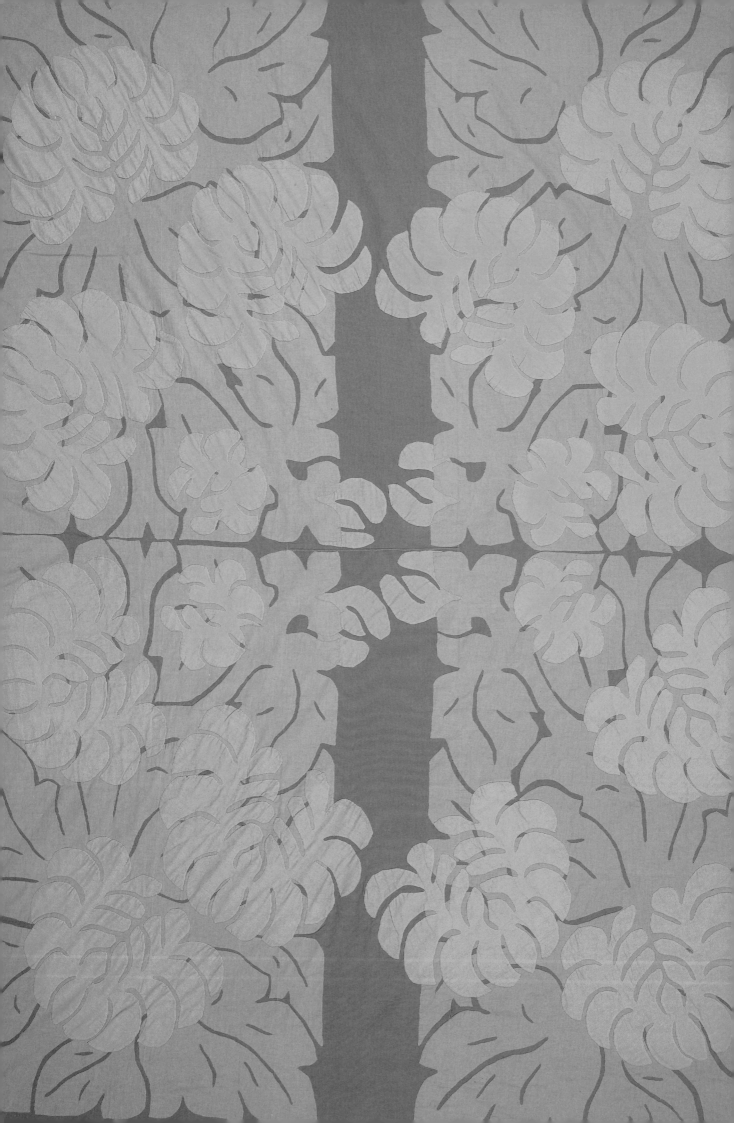

Chapter 3

Keeping time, stitching memories

Patchwork quilts and the mapping of time and space

The association of the patchwork quilt with memory, with protection and with the sustaining of connections hidden from plain view has inspired generations of scholars and practitioners. The quilt frequently inspires nostalgia. A tangible thing that has been crafted, altered and used by people, it has a capacity to tell of times gone past possibly only matched by narrative.

As Judy Elsley has argued in her book on *Quilts as Text(iles)*, the patchwork quilt can be compared to a text which takes shape in blocks that form an ever larger pattern 'whose central motif is change' (1996: 1). While the analogy with writing is not one I want to pursue in this chapter, it is useful to us here in as much as it allows us to discern a unique feature of the activity of quilting: patterning a surface, like a pen on paper, the pieces which compose the quilt serve to draw ideas together in a manner that speaks of a synergy of making and knowing that continues beyond the construction of the 'final' product. In the same way as we readily accept that the writing down of a narrative makes its content portable as well as comparable and combinable with other narratives written before and after, patchwork made of stitched fabric pieces lifts the resulting pattern onto a spatial and temporal plane where it functions no longer as a singularity, but as something that combines and multiplies, a provocateur and agent of transmission.

Patchwork as a technique has its roots in the European decorative sewing used to embellish clothing, bedding and other textile items, either singly, or in combination with quilting, which, in fact, is a different process altogether (see Horton 2005: 17; Schoeser 2003: 112–33). The two processes may be combined, as they have been in American colonial history, yet each can also be used on their own to create planar surfaces in which patterns appear through stitching. In quilting, two or more textiles are layered through stitches that penetrate all layers and form designs through the alignment of their rows. Patchwork, on the other hand, involves combining small pieces of fabric to create a surface pattern; this additive process may either take the form of piecework, in which fabric pieces are sewn directly onto each other, or of appliqué, in which fabric pieces are arranged and sewn onto a fabric background that is part of the pattern.

Where unquilted appliqué occurs on its own, superimposed onto a second layer of fabric, as is the case

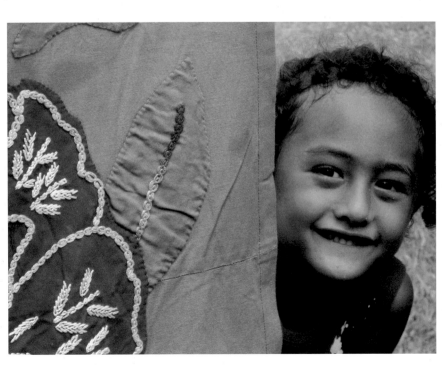

Left: *tivaivai manu* of chrysanthemum flowers designed by Ake Takaiti, sewn by Ake Mingi, Atiu. 216 x 275 cm (detail). British Museum Oc,2003,02.2.a.

Right: Connections hidden from plain view.

Tivaivai versus tivaevae

A battle has been going on between the outer islands and Rarotonga as to whether the term for the colourful coverlets that are the subject of this chapter should be written with 'ae' or 'ai'.

I had not yet seen the word in its written form when I asked the Atiu *mamas* how it should be spelled. Their response was: 't–i–v–a–i–v–a–i'. But then a Rarotongan friend told me that I had written it incorrectly, and that it was spelled with 'e'. For many years I regarded this as logical, for when you look the word up in either of the two Cook Islands dictionaries (Buse and Taringa 1996; Savage 1980), they inform you that '*tivae*' means 'to patch, to mend by putting on or inserting a piece', and from this *tivaevae* translates as 'patchwork, [...] a name given to a bedspread or cover made up of patches of material'.

Only when I started my research did I start to ponder the origins of this orthographic battle. Surely the problems that native English speakers had, and still have, with reading and writing a phonetic language provide too glib an excuse for the ambiguity. I would have to look at the use of *tivaivai* and the role it has played in Cook Islands culture if I wanted to understand the implications of the different namings.

When considering the coverlet as mere decorative object, it is clearly correct, in view of the techniques used, to write '*tivaevae*' and to translate this into English as 'patchwork'. However, when considering its use, which so closely connects it to the ceremonies and customs requiring barkcloth in pre-missionary times, '*tivaivai*' may be just as appropriate. The coverlets came to the Cook Islands from Tahiti, where they were and still are called *tifaifai*. '"*Tī*" is an intensifying prefix with the force of an adverb [that] gives distributive force to verbs, etc.' (Savage 1980: 376). It carries the sense of 'wholly', 'entirely' or 'all over'. The word *vaī* means 'to wrap up, to enclose with or in something, to encase or fold together, to enfold' (Savage 1980: 448). Surely it would therefore make sense if '*tivaivai*' were the word used to describe a ceremonial 'cover' or 'wrapping cloth'? It certainly was, and is still, used as such.

in Eastern Polynesia, the resulting artefact is fragile and impractical for daily use, as any washing would be likely to shrink top and bottom layer in different ways. There is often a confusion around Eastern Polynesian quilts, which actually fall into two types. In Hawaii, where patchwork was introduced by New England missionaries, three-layer quilts resulted. In the Society and Cook Islands, however, the teachers of patchwork originated in Europe, where batting and quilting were unfashionable in the early nineteenth century. Here, notably in the Cook Islands, patchwork took the form of unquilted coverlets. The term *tivaivai*, which refers to such coverlets, elicits associations with *vaī* (to wrap up, to encase), *va'i* (to break, to split, to cut), *vava'i* (to split lengthwise), *vavai* (cotton), *vai* (to be, to exist, to occupy a place, to remain) and *tivae* (to patch, to mend).

The fragility of *tivaivai* is heightened by their size and by the type of cloth used. The latter is not colour-fast, and its suitability has more to do with a mode of fabrication that involves tearing cloth into pieces, than with durability. Were one to use a piecework *tivaivai* as a bedcover, the time invested in the stitching of the tiny pieces into place would be in sharp contrast to the brevity of use it would afford; indeed, as Horton suggests, even in the nineteenth century, utilitarian considerations would have steered busy American frontier women away from the patchwork quilt as 'impractical, unattractive, and a waste of time' (2005: 18). Patchwork must thus be motivated by considerations that lie beyond any utilitarian appreciation, the kind of value the craft status of the quilt might suggest.

In fact, we get a clearer sense of the motivations behind patchwork when we ask what the sewing of fabric docs 'do' if it does *not* construct items for everyday use. The activity's utility is demonstrated in studies, most from America, that examine entire collections of quilts, and their patterns, to shed light on relations of kinship that have connected women to each other and to the homes they have inhabited (Burdick 1988; O'Bagy Davis 1990; Marston and Cunningham 1990; Snyder 1993; Horton 2005). Patchwork appears to encode in its patterns a map of social relations that animates genealogical information by allowing the quilt to stand in for the maker and, occasionally, the person it was made for. The resulting fabric represents experiences and connections that can only be recalled from memory, rather than any document or record, enticing us to reconstruct the social field within which it would have been made and received as a meaningful product of work undertaken within the home.

Beyond genealogical information, quilted patterns have been shown to document a microscopic perspective on attitudes to material culture within the home, which can be contrasted with the macroscopic perspective that is generally captured by historical studies of wider societal patterns of consumption and identity construction. Patchwork allows us to get a glimpse of preferences for texture and colour, and even the print or the smell of cloth, its weight and the sound created when unfolded, all of which say as much about the temporal and spatial structure of the household as the pattern created through patching pieces of cloth together.

Patchwork makes tangible the wider material relations that encapsulate the character of a place and a time, and can allow us to sense these relations long after they have perished. The patchwork draws one into its own world by virtue of the directness with which it represents the lived-in material world, presenting it to us as a retention, a recollection, that we can then use to think with when comparing here and there, now and then, and when plotting the future course of our own lives. Yet, more than anything, it is pattern that attracts attention, suggestive as it is of the maker in ways that are difficult to fathom. There is the stitch, the size and shape of the pieces of cloth, and the preference for particular ways of aligning them into a regular pattern; yet we may find it difficult to see how the resulting product can stand in for a person who may no longer be around. Patchwork, in fact, appears to be a means of drawing fragments of a person's biography together in ways analogous to polyphony – a tightly structured composition that schematizes, and thus enables the recall of, biographical time, itself made up of manifold relations coming together in ways that can only be appreciated retrospectively.

Patchwork, however, does not just invite us to an insider's perspective on local history, but also tells us about the wider economic and cultural conditions that provoke women to harness and transmit cloth wealth by cutting up and reconstituting material into planar surfaces in ways that invite such acute retrospection. It may be no accident that biographical studies based on the material culture of the home have been made most successfully using American quilt collections from the nineteenth century. The prevalence of patchwork quilting in the culture of colonial America tells us much about the heightened status of fabric resources in the domestic economy run by frontier women, which was itself motivated by the rise of mass-produced printed cotton made possible by innovations in the British textile industry; and also about the increasing relevance of textiles in international trade since the early nineteenth century (Mukerji 1983: 218–28). A new attitude toward thrift, which became dominant in America with a rising consumer culture in the late 1900s,

shows up in the reuse of dress material. American quilts of this period are a record of the motivations, sensations and reactions that women associated with printed fabric, which told of an unprecedented transformation of private and public lives punctuated by fabric innovations in clothing and home furnishing (Sennett 1974: 64–88); by cutting up and re-stitching those fabrics which had been worn on the body or bought for display in the home, women can be seen to have both participated in that process of material innovation and to have consciously preserved its trace for future generations as a record of 'living history'.

We can thus assert that patchwork, wherever it is made, is likely to possess a distinct historicizing potential, as it serves as an instrument for the conversion of value attached to materials and commodities, such as fabric, into intangible values attached to the moral economy of the household. The success of such a conversion is inherently linked to the importance of the quilt as agent of transmission; and the fact that a patchwork is made of fragments of cloth is more than a metaphoric allusion to the quilt's capacity to 'gather' the bits and pieces of one's life into a single frame, very much like a written biography.

The prevalence of patchwork in households, therefore, does not speak so much of economic necessity or virtue, but of a value placed on fabric as a material that is uniquely able

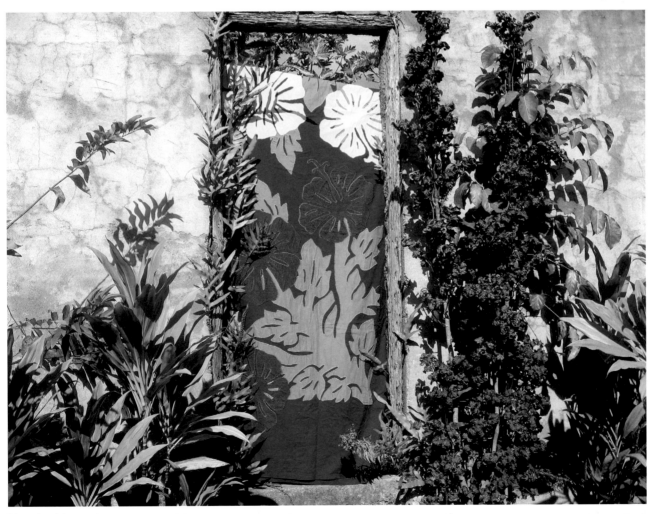

Memories of times past.

to blend together the tangible and intangible resources which together make up a person's identity. Patchwork quilts, the world over, are stored and treasured, rather than sold and bought, and hold their value over time as a result of their ability to personalize what otherwise may appear as generic cloth wealth. The reason for the patchworks' apparent capability to transcend the main manner by which things enter into sociality, which is by binding value into patterns of consumption, is, however, unlikely to be explained just by their materiality. Their fabric does not adequately distinguish the patched quilt from other forms of cloth wealth, which as a general category may encompass a host of fibre-based products, such as weaving or clothing. Rather, it is to the techniques of assemblage that we must look in seeking to understand how patchwork can convert cloth into a measure that can be used to gauge and reckon relations that exist between persons and between persons and things, and thereby into a matrix that can define and retain the identity of a person in a material frame.

The patchwork quilt already denotes in its name the laborious task of 'patching' that captures the success and effectiveness of the household economy; the fragility of the patchwork is a tangible reminder of the work that is literally demanded by the fabric for it to be of value. As Horton reports from her research into colonial American probate records, 'quilts were less common and more highly valued than blankets, woven coverlets or bed rugs'. Describing the location of quilts among the fashionable families of Charles Town, she notes: 'throughout the eighteenth century, in both Europe and the Colonies, quilts were made from fine fabrics, using silk or imported cotton cloth. Quilts of this period represented a considerable investment of time and effort and were generally made and used by upper and middle-class families…' (Horton 2005: 17).

Patchwork is thus indicative of a surplus of fabric, but also of a surplus of time – suggesting a division of labour within and between households that enables the conversion of a tangible commodity into a manifestation of an intangible resource; when quilts are sold or gifted in life-cycle events, their value is not measured in terms of the original cost of the fabric used for the manufacture, but in terms of a monetary value placed on the time it took to sew. Because it is not so much the material value of the fabric as the embedding of time into its folds which matters when patchworks come to market, such patched quilts may be best described as a 'sewing' – a notion which expresses the temporal performativity of patchwork, whose material life is punctuated by moments of airing, inspecting, unpicking and re-stitching.

What is that time, however, which is thus measured by the technique of patching? Is it an abstract notion of time, a quantifiable entity of fragmented durations of a specified kind? Or is it an experienced measure of time, a moment in a woman's life that may be as long or as short as it appears in recollection? Both notions of time are grounded in experience and may be elicited by the patchwork quilt, yet neither tells the full story, because time itself is transformed into a pattern in sewing patchwork – a pattern which demands a certain apportioning of time.

There is perhaps no better example of the difference made by patterning to the way we think about the time that is measured by work than Cook Islands *tivaivai*. There are three types of technique found in Cook Islands sewings, each with its own characteristic *durée*, which itself resonates in complex ways with the temporal nature of the biographical relations people draw on when making claims to identity, land, and knowledge. As we try to find answers to how these techniques are distinguished from one another and why they have

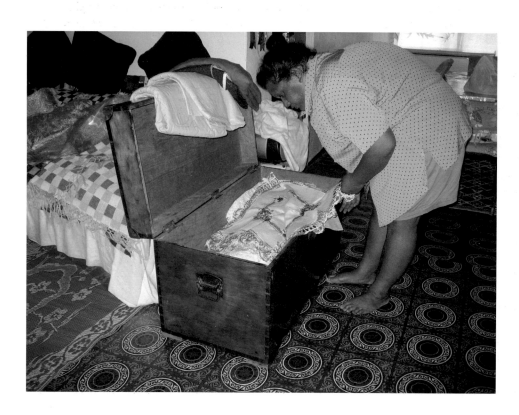

Tivaivai are stored folded in large treasure boxes.

Fabric at 'Vonnia's Store'.

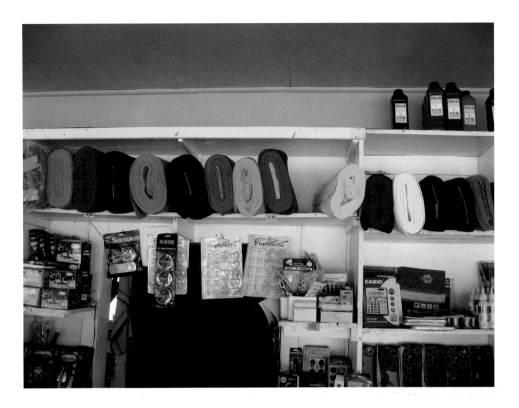

persisted over time, we will discover that in the Cook Islands patching works in a way that transcends the retrospective perspective so apparent in American patchwork, and is in fact central to socially established approaches to learning from past events when considering one's future.

We may think we are unaccustomed to the possibility of mapping one's life-course in a manner that broadly corresponds to a way of envisioning trajectories of influence. In fact, while lives in Europe and America are undoubtedly marked by increasing individuation, they are also affected by an increasing emphasis on the role of statistics in forecasting the future and steering personal and societal decision-making (Hacking 1990). In the Cook Islands of today, statistics reign just as firmly in the making of societal decisions, but where the welfare and life-course of its people are concerned, we find that quilts map the contingent relations between past, present and future just as effectively. In fact we could argue that patchwork has become prevalent across Eastern Polynesia because of its ability to measure and map time in ways that were rather inhibited by fibre-based technologies such as mat-making or the beating and decoration of barkcloth (c.f. Küchler 2005). We shall see that patched patterns can encode memories of life-changing events just as effectively as they can map the logic of biographical relations, drawing them together at the very final moment of life, when the quilts that have marked the relations which encompassed a person's life are wrapped around the body in death.

Patchwork is hugely popular across Eastern Polynesia, and is essential for the ceremonies that punctuate the life-cycle, from significant birthdays and first haircuts to weddings and funerals. Possibly as a result of its role in cementing and crafting the transnational relations that comprise Cook Islands identity today, patchwork sewing is now utilizing

'new' material, thereby detaching the materiality of the quilt from the associations that worn and reused fabric harbour. As sewing in the Cooks uses store-bought cloth, and as quilts are rarely stored and circulated in exchanges for longer than a generation, the retrospection that the patchwork quilt invites so clearly in America is instead turned into a prospective enterprise. Numerous events requiring new cloth and new sewing fill household calendars in the Cook Islands and its many diasporic offshoots.

We cannot understand the resonances provoked by such sewing unless we note that its pattern is not the result of a random assemblage of scraps of cloth, but one that documents a consciously executed plan that started prior to the purchase of the material; a pattern, moreover, whose complex process of creation expands the functionality of the sewing beyond that of a decorative cover, and situates it squarely into matters of direction-finding and future-oriented thinking. The Polynesian quilt, and the Cook Islands sewing in particular, is thus far more than the documentation of something that we might verify externally by recourse to history, oral narrative or written or remembered genealogies; rather, it is the means by which such discourses about history and biographical memory are made possible through recourse to an abstract map of the biographical relations that shape people's life courses.

African American quilts from the time of slavery are an example of quilts whose patterns in fabric direct the mind to new places and potential futures. These eschatological quilts are a poignant reminder that quilted patterns are capable of inspiring a vision of a possible future, as they were an instrumental part of the struggle for freedom from slavery. By means of these quilts enslaved blacks travelled along the 'underground railroad', which was stitched into the fabric as

A proud job

In the early days of our studio a friend from Rarotonga commissioned us to sew a *tivaivai taorei*. When his grandmother died, he had found an old suitcase full of unfinished parts of patchwork and one whole *tivaivai*, without backing. Unfortunately, the *tivaivai* was in very bad shape and could not be restored to its former beauty, parts of it having fallen prey to insects and mould. He remembered seeing his granny sew this particular *tivaivai* when he was very little. As his daughter's wedding was near, he had the idea of commissioning us to make a replica of this old heirloom piece as a gift for his daughter on her wedding day.

For me this was a welcome challenge. No-one on the island made *tivaivai taorei* any more. I wanted us to do it the really traditional way, as we might be the only women of our generation to sew one; we would then be in a position to pass on our knowledge to future generations once the old *mamas* were gone.

In those days, special fabrics were hard for us to get. Telephone had just been installed on the island; previously communication with the main island or overseas had been by radio. I took advantage of an overseas trip to buy the special muslin fabric needed for our commission. I was able to match most colours except for the pink: the one we ended up using was rather dark.

Tini, Teremoana and I took an entire week to rip, cut and string-up the pieces in their correct order. The total pattern consisted of 27,500 one-inch squares! When we each finished our first segments, a triangular-shaped eighth, we held them next to each other. Much to our dismay, we discovered that each one of us had used a slightly

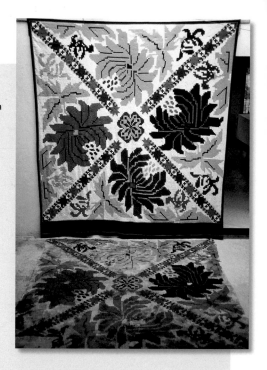

Original and replica of an old *tivaivai*.

different seam allowance, which, multiplied by 3,437 patches, made a visible difference. We decided that Tini should finish the remaining eighths, because her triangle was the middle size. Some 500 hours later her job was done.

We sewed the pieces together and stretched the whole *tivaivai* on our working table. With the help of drawing pins, a steam iron and some patience, the work ended up being remarkably square and one could hardly spot the two different-size eighths. When we sewed on a backing, we also slightly enlarged the *tivaivai* to become rectangular, so as to fit our client's bed size. Seeing the original and replica together almost resembled a reflection in the water. A proud job well done!

---

an interconnected web of points that could be mapped onto the landscape (Tobin and Dobard 1999). These quilts, serving to transmit the paths to be taken into freedom, show that patchwork may surface as the main vehicle of communication in situations of dislocation and dispersal that foreshadow the fragile connections of a community in diaspora.

That quilts may be maps may strike us as odd as the idea that they are instruments for the conversion of value, and thus a superior resource within a household economy. Maps of any kind are part of a cultural system of way-finding, in which spatial and temporal acts of making and tracing lines lead beyond the visual field of perception to the places of one's dreams. Way-finding tools can take diverse forms suited to differing modes of travel: the demarcation of points and overlapping fields of influence is particularly suitable to seafaring, something that throws a new light on the speed and zeal with which needlecraft, tracing with its movement the fields that connect points in space and time, was taken up in Eastern Polynesia, where small islands are connected and divided from each other by fields of currents (Ingold 2007). By comparison, in the western part of Polynesia and island Melanesia, where land mass is more prevalent, the

hand-stitched cloth that is synonymous with patchwork or embroidery has been superseded by the sewing machine and ready-made clothing bought at the market. It is possible to generalize from the Cook Islands, to say that sewing has enabled women to take up the previously male skills of navigating across Eastern Polynesia, both in the literal sense of spatial orientation across a massive seascape and in the temporal sense of transmitting land-rights anchored in genealogy. Armoured with maps of their own, women have migrated from the Cook Islands and the other nations of Eastern Polynesia in greater numbers than men since the mid-twentieth century, moving long- and short-term to the metropolises of the Pacific and beyond.

Cook Islands patchwork will thus be presented here as a technology of mapping that is particularly effective at drawing together a discontinuous social world composed of multiple nodes and stoppages. We shall see how the patterns that emerge on the surface of cloth through the act of sewing thus carry clues to the means by which a multi-centred community sustains a sense of gathering and thus of belonging (Lippard 1997).

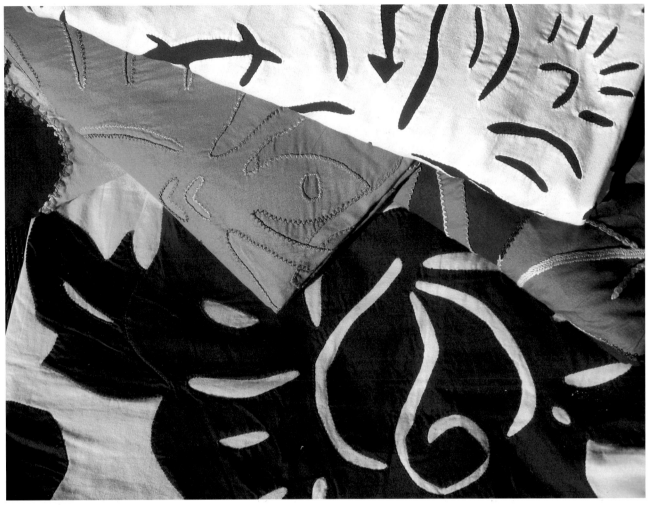

Manifold connections.

What stitched maps represent

Although African American quilts were made for travelling across land, rather than the open ocean, they too transmitted spatial concepts and pathways in codified form. The songs which accompanied the making and displaying of quilts in the hands of slaves would clearly have aided the translation of the topology of the quilt into a de-centred conception of space that enabled people to 'see' the 'underground railroad' as they navigated through unknown territory into freedom.

To visualize how quilts may enable a translation from the mundane representation of space as a kind of container or box which is filled with something we call matter into the much more difficult to comprehend notion of a continuous surface, and back again, it is useful to recall how quilts are made. In sewing it is impossible to arrive at a product that displays a modicum of regularity without advance preparation and planning. Decisions need to be made about the type and quantity of material, the shape or shapes it is to be cut into, and the order in which to assemble the pieces of fabric. When starting to work on the quilt, the needle threads the pieces together, with the eye following the course of the needle in the same way as it will traverse the surface of the finished work, taking a course that will effect what we may call the

recognition of its pattern. Looking at this planar surface, or anticipating it in production, utilizes mental images that are derived from our experience of space, which is of course shaped by culture as well as being something that adheres to certain cognitive universals given by our predisposition to experience space walking upright in a directional manner that is limited by the material environment around us (see Johnson 1987). The translation of a mental image of space bounded by things into an image in which space is limitless, teasing the eye to move from point to point to search for connections between visual fields that cluster around points, is a task we call mapping.

Mapping, of course, will lead to different kinds of maps, depending on what kinds of preconceptions we may have about the nature of the connections between the points. A medieval map of the world, for example, placed emphasis on the biblical connections between points, thus creating a hierarchical map of continents and places radiating outward from the point of creation. Although it does not correspond to our experience of the world today, we can still 'read' this map with the knowledge of the cultural conventions that pertained at the time of making. The fact that we are likely to miss the point that quilts are representations of *space-time* can be explained by the influence of culture upon the selection and arrangement of images that represent the world as it is experienced and perceived. However, this answer just begs the question of what the 'cultural' element in the shaping of such images of space-time consists of.

Much has been written in psychology and anthropology about the conception of space, the representation of which appears to fall into two types: point–field and automatically bounded 'containers'.[1] The latter is the most widespread schema, and is used the world over. The lack of an explicit name for this representation of space as a box filled with matter, despite our colloquial use of it (people talk about space as if it were a 'place something is located in'), may indicate that it serves

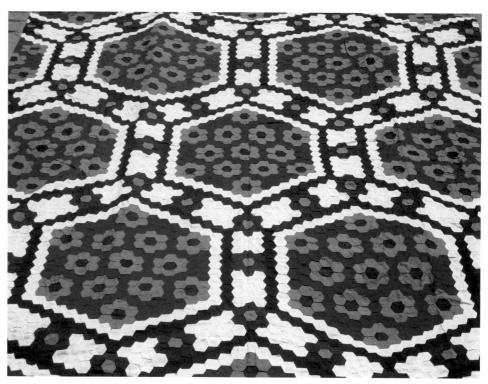

Tivaivai taorei (*paka onu* - hexagons) owned by Kimiora Samuela.

as the default mode of spatial thinking. It is only the advent of modern, relativistically oriented physics and cosmology that has seen the introduction of a point–field manner of representing space as essentially *relations between points*, with distance being derived from their interrelation *through time*.

In its attempts at understanding the nature of non-linear and topological space-time in which boundaries are defined by the symmetrical overlap of adjacent topological fields rather than as existing as axiomatic constituents, contemporary physics directs us to an important, perhaps rather obvious, thought, that in the lived world, time and space cannot be disentangled. The connection between spatial temporal conceptions, however obvious it may be, has only sometimes been understood by anthropologists, notably by Granet (1968) in his important account of Chinese formulations of time and space. Granet's study of Chinese conceptions highlights the role that temporal conceptions play in mapping the world, the understanding of which had enabled the Chinese to navigate oceans more effectively than the Western powers. In fact, they had arrived at a way of correlating units of space with standard units of time, and of measuring and recording them accurately. One wonders about the material supports for the conceptual and mathematical skills enabling Chinese navigation, but it may be no coincidence that Chinese sailors operating aboard ship in and around Hawaii in the eighteenth century were known to have practised patchwork, although we do not know what kind of associations

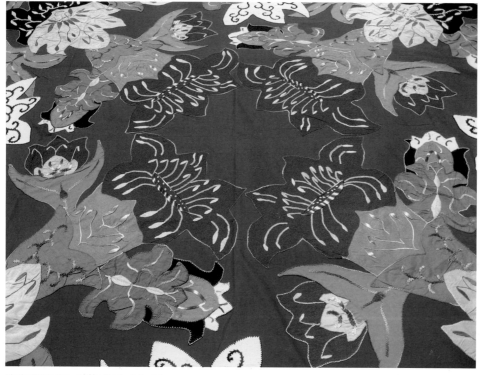

Tivaivai tataura owned by Ngakura Kautai.

existed between this activity and the accounting of time. We take the preoccupation with keeping the 'correct' time for granted in much of our daily lives today, but in fact the discovery of how to keep time accurately is a rather recent discovery, and it was one which was instrumental to the technological change and economic development of Europe after 1750 (Landes 1983). The clock, and the keeping of time, has subsequently been central to commerce and navigation in the West.

However, the preoccupation with keeping time in a linear and cumulative manner, exemplified by the clock, is not always found in cultures (such as those of the South Pacific) where livelihoods depend upon successful navigation. A famous example of skilled navigation using maps of a kind that do not require the element of linear time, but which instead imply fields of relative temporal motion, is the 'stick-chart' of Micronesian islanders (Ascher 2002). Such charts are geometric arrangements of sticks and shells that depict the interplay of oceanographic phenomena such as wind and sea interaction in and around the atolls, as well as the land masses. Using a relational and analogical mode of representation that depicts the interaction of shapes (land masses) and motions in the sea and at the sea–land interface, the islanders deploy both idealized maps for teaching and 'working models' for navigation around such complex topography, and even distinguish features of the underwater landscape.

Time is depicted in images of constricting or expanding motion between connected points, which thus describe space in de-centred and relational, rather than relative or egocentric, terms. Jürg Wassmann (1994) has pointed out that Pacific artefacts such as knotted cords, like the stick-charts of Micronesian seafarers, direct attention to a topological conception of space that appears to prevail in the ocean-dominated landscape of the Pacific. Nowhere else may the practical significance of mapping topological relations of space-time be more straightforwardly important to people as in the Pacific, where a constellation of island communities are dotted across an ocean expanse covering over one-third of the world's surface area on a total land mass that, once Australia is excluded, equates to less than one-eighth of that of Europe. The Cook Islands are in fact a typical Pacific island community, with its group of fifteen tiny islands and atolls, of which thirteen are inhabited (sometimes by only a few

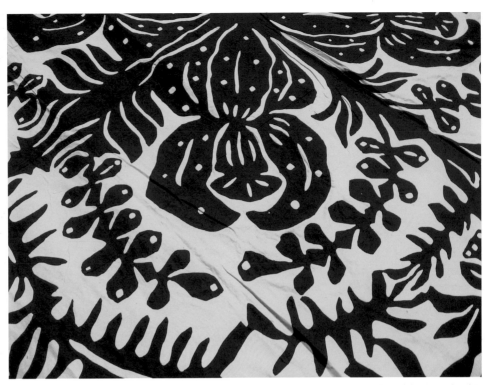

Tivaivai manu designed, cut and sewn by Krudean Kareroa.

hundred people), dotted across an area the size of Europe. Despite their lack of proximity, the people who inhabit these island communities are culturally and linguistically linked, and are known to have had longstanding and extensive trade and kinship connections with other even more remote island communities, such as the Society Islands some 2,000 miles to the north-east. In this environment, water is the predominant feature and resource, and is a pathway along which people, their produce and ideas travel.

In a world dominated by the ocean, it is thus not surprising that mapping and artefacts of navigation have attracted a good deal of attention from anthropology and related disciplines (Hutchins 1983; Gell 1985; Silverman 1998; Bennardo 2002). What is surprising, however, is that the importance of sewing in the Pacific has so far been discussed in literature without paying any attention to spatial and temporal conceptions and their representation within an ocean-dominated landscape. This is perhaps because quilts fall so awkwardly between categories of things, between the useful and the valuable, that it is difficult to see what to do with them other than treating them as what they appear to us to be in a rather straightforward manner: proof that time was expended that was, in a way, left over as 'free' time to devote to oneself and one's family. And yet, it is obvious that it is precisely this observation which must force us to look closer at what is done with this time, how this time is made materially graspable, clad in images that fill the empty hours of the day as much as they fill the gaps between the points at which the needle is inserted in the cloth. Time is made tangible through sewing as it surfaces in images that capture its continuous, directional and repetitive nature. It is by attending to these images as culturally specific mappings of biographical relations that exist in 'real time' that Cook Islanders are able to navigate their life through the fractured social and spatial landscape of the Pacific.

The mechanics of time-keeping: clocks and quilts

No historical data document the take-up of sewing or the fashioning of patchwork across the island nations of Eastern Polynesia, from Tonga and the Cook Islands to Tahiti and Hawaii. For the Cook Islands we do know that cotton cloth was a sought-after trade item, and that cotton planting and printing were well-established on the main island around 1850, barely 25 years after the London Missionary Society had arrived on the islands (Gilson 1980: 44–7). Although many interpretations of patchwork in the Pacific have concluded from the value ascribed to cotton fabric in the heyday of missionary activity that the sewing of quilt-like fabrics was learned from missionary wives, an alternative and interesting hypothesis advanced by Robert Shaw (1996) argues that the inspiration

for patchwork quilting happened prior to, and independently of, missionary activity in the Pacific. Writing about the history of the Hawaiian quilt, Shaw questions the link between patching, which was indeed taught by missionary wives to foster domestic attitudes and time management, and appliqué, which in Hawaii resulted in a quilt that involved the cutting of cloth into snow-flake patterns (Akana 1986).

Noting the absence of this type of sewing in America during the early part of the nineteenth century, with the exception of Pennsylvanian missions who taught German paper-cutting techniques at mission schools but only set foot in Hawaii in 1860, Shaw points out that Hawaiians may have been familiar with sewing, clothing and even quilts from contact with the sailors on the trading ships that preceded missionary influence in the Pacific. In 1800 you could already buy almost everything on the shores of Hawaii, as the booming trade in sandalwood had brought Chinese silks,

The fabric of friendship

She bends down low to retrieve a large metal tin from underneath her cast-iron bed. Then Mami Tepu sits down and pats the bed next to her. 'Come,' she invites me with a smile. When she takes the lid off the tin, fireworks of colours explode in my eyes. 'This is how we make a *tivaivai taorei*,' the old lady explains, as she pulls out a half-finished triangle composed of thousands of tiny fabric squares in a rose pattern and something that looks, to me, like a caterpillar: fabric squares strung up in sequence. And teacher and student sit together side-by-side until the sun sets and it is time to prepare dinner.

That day I learned how to make a *tivaivai taorei*. But what I learned was much more than sewing. With each fabric square that I moved from the string I received a glimpse into a Cook Islands woman's life, so different from the lives of European women, and yet having many underlying similarities. And as I sewed each square to its tiny neighbour, I pieced together another section of the large puzzle that life on this remote island has posed for me since we arrived nearly a quarter of a century ago.

Mama Teau Kea, who we all called Mami Tepu, was the wife of Parua Ariki, one of the three high chiefs of Enuamanu or Atiu island. When the idea to establish the Atiu Fibre Arts Studio was born, she was the first person I asked to become one of the company's shareholders. She happily accepted and was elected one of our company's directors. But we had to postpone the opening of our studio nearly a year, because shortly after I approached her, Parua Ariki died.

Over the years, she and I sat together many hours as teacher and student, sometimes reversing our roles. With even and firm stitches, we constructed a friendship that spanned across the hemispheres of the world, across a generational gap (Mami Tepu was twenty-five years my senior) and across differences of cultural background that couldn't be more opposed. Sometimes we felt like mother and daughter, and it felt good to have a mother near me, when my own was on the other side

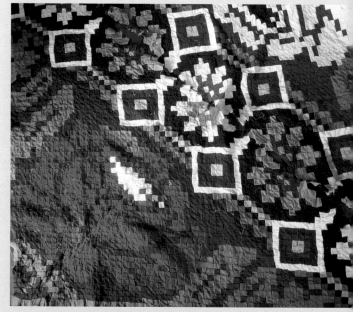

Mami Tepu's unfinished *tivaivai taorei* (detail).

of the world. Sometimes we were just two women who had childlessness in common, and who shared many other experiences that weave through the lives of all women like the thread that our hands stitched through this fabric of friendship.

When Mami Tepu died, she was buried next to our house in her family's graveyard; she has remained the mother that I can visit when I need her advice.

The house that used to be her home has stayed empty. Her clothes and some *tivaivai* were buried with her. Before she went on her last journey, she had packed bags for her adopted children containing *tivaivai* and other sewings she had made for each one of them. Family members cleaned out the house. What had not gone to the grave or been distributed amongst the children and other relatives was burned.

cotton and costumes, transported by scores of Russian and Chinese ships. Though there is no literature to substantiate a link with these ships and the take up of patchwork in the Pacific Islands, we do know from maritime collections that the products of sailors' long hours of 'idleness' at sea were not limited to engraved whale teeth (scrimshaw), but also included macramé (a type of lace) and patchwork. Indeed, a link seems likely between the sailing ships' extensive use of cordage, not least to bind together the technical means of controlling motion with the ship's material frame, and the idea of their being an appropriate place for sewing, a link that can equally be made with the construction of canoes.

The idea that sewing enables the literal 'stitching up' of time, by giving it a material frame from which to comprehend and manage time that is experienced as discontinuous, with landmarks that punctuate its flow, is a powerful one in the Pacific, where the trading ships' voyages between islands

would have lasted weeks at a time. Still, today, the time it takes to move from point to point across the ocean is one that is felt deeply by islanders, who in bad weather often wait weeks for ships to bring fresh loads of flour or petrol. The desire to make time connect more effectively with points beyond the horizon may thus quite feasibly have guided both Western sailors and Pacific Islanders towards the manipulation of cloth in a manner that came to take on the form of patchwork (Siikala 1991: 3).

Much of the organization of the home introduced during the nineteenth century by the London Missionary Society clergymen and their wives as part of their effort to inculcate new ways of thinking and being still exists today. The household is a microcosm and foundation of the Cook Islands' economy, but to imagine isolated units striving to maintain a balance sheet of work and leisure would be entirely wrong. The social organization of production and consumption connects

---

One of my friends came home some weeks later and gave me a Hawaiian quilt book that she had rescued from being thrown on the fire. All of us shareholders had received the same volume from a Hawaiian friend when he visited Atiu, except for one *mama* who had not yet been a shareholder in our company. We decided that she should have it. That was in 1994.

Some thirteen years later, two of my friends visit me one day with a plastic bag containing fragments of a *tivaivai* they want to show me. When the cloth wrapping it is removed, I feel like I am meeting a long forgotten friend. Memories of those happy hours as Mami Tepu's sewing student come back instantly, as I look at the rose pattern before me. I can see that these pieces have been washed, because the pink has run into some of the white squares. Well-meaning family members had also rescued this unfinished *tivaivai*, and my friend's sister asked if she could keep and finish it. I am so happy that this *tivaivai* has found hands that will finish piecing the puzzle that Mami Tepu and I began twenty-two years ago.

But that is not yet where this story ends. It is close to Christmas, the time when relatives who live overseas come home to the islands in great numbers to visit family and friends. I have just started to work on this book project, and I ring my friend to ask if I can borrow her beautiful *tivaivai manu* for photographs. Five minutes later I hear her motorbike. She leaves her *tivaivai* bundle on the table, and we chat for a while. 'I've got to go. I'm just cleaning my home for the visitors,' my friend then says. I accompany her back to the scooter. On the seat is another bundle of cloth that she picks up and holds out to me.

I found an old *tivaivai* that was rescued from the fire after Mami Tepu died. The backing is really rotten, but you just need to take it off and replace it. At home it is in my way and will only rot away further. I think that you should

have it, because you are so passionate about those things. I know you will be able to look after it. Mami Tepu would have wanted it that way.

She smiles and is gone before I have recovered from the surprise.

She is right; the backing is quite rotten, ripped to shreds, faded from what was once a rich emerald green to muddy beige, and it shows signs of mildew. But the *tivaivai* itself is still beautiful! Though partly faded and torn, once the disgusting backing has been removed the yellow-and-orange-butterfly pattern radiates its former charm. Especially on the reverse, now revealed, the colours which had been hidden from light by the backing fabric still shine with their original happy brightness.

To me, this is like a message from beyond, an embrace from my motherly friend. Perhaps this is the *tivaivai* that will one day accompany me to my grave when my time comes to move to the family graveyard next to our house?

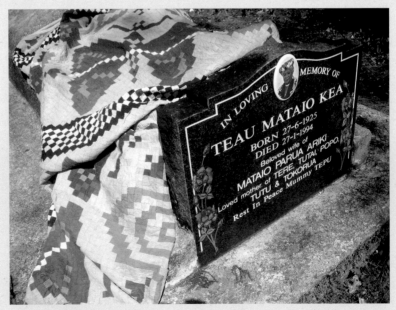

Mami Tepu's gift from the grave.

households in the most effective manner possible. These connections, which bind households together socially as well as economically, are founded upon sewing bees in which women participate on grounds of friendship rather than through genealogical connection. Such friendships may develop in and around activities organized in relation to the church, but they may also be based on time together in employment or in running communal affairs such as girl guides. Memories of particular events or relations may prompt the creation of patterns that are forever after associated with moments that mark women's biographies. One-off encounters and events such as collective visits by church groups to Australia may prompt new sewings, just as kin-based events such as the first haircut of a son, a wedding or a funeral may do. The biographical relations and memories that are created in the vicinity of patchwork are today given huge prominence, over and above the genealogical connections that connect people to their tribe and to the land that they are allowed to inhabit and work upon.

The patchwork carried out by such sewing bees punctuates the hectic yearly calendar of every woman: there is the date set by the minister's wife for the annual or biennial competitive display of sewing; the group trips to workshops or visits, each bringing the need to carry sewing as a gift; as well as the many household-specific ceremonies marking birthdays, weddings, funerals and commemorative events. Women's sewing groups usually have a joint bank account, stocked regularly from the sale of sewing at exhibitions and other sources of income, to finance the purchase of new material, but also to update the contents of each other's homes: taking it in turn to put their husbands to shame by purchasing a new cooker, a new set of dishes or whatever may need replacing or adding to a house.

Managing the time for one's sewing work is of great importance, in order to fit in with the calendar of competitions and ceremonies that punctuate life in the Cook Islands. Finding time for sewing is not always easy, as many women also have to attend to paid employment, gardening and participation in the many year-round village sports, as well as their daily chores. Some women have never found the time or opportunity to learn to sew the fine stitches required for the making of quilts in the Cook Islands. So many women resort to employing others, who may be expert cutters of cloth or particularly good at stitching a fine and regular line. Younger women who find living up to their aspirations difficult like to recall the feats of their elder sisters, mothers and grandmothers, who managed to 'do it all'.

The time of sewing also constitutes the value of each quilt, as well as its labour cost. Women can estimate with ease the expense incurred in purchasing each other's time as well as skill. Several women have reciprocal arrangements to gift each other with time, which is in any case a prerequisite for the collaborative sewing of a large patchwork quilt, whose triangular sections are stitched by different women, who expect as remuneration the return of such favours when it comes to the construction of their own quilt. By joining cooperative ventures resulting in a patchwork quilt, women also gain access to the pattern, which forever is associated with the circle of women and the circumstances that brought them together. This capacity of patchwork quilts to connect those who are of different generations, exemplified by the gifting of quilts from grandmother to granddaughter or grandson, makes these quilts the most valuable of all, as measured literally by the time retained in their patterns.

There are three types of *tivaivai* in the Cook Islands. The less time they take to make, the shorter the biographical distance they are gifted across. The most arduous of all quilts is a type of patchwork that should strictly be called 'piecework' (*taorei*) to emphasize its construction from several thousand tiny pieces of cloth that need to be joined together into a regular, symmetrical and self-replicated pattern. This is often gifted by a mother to her daughter at marriage to be wrapped around the daughter's husband in death, or by a grandmother to her favourite granddaughter or grandson. Today such *tivaivai* have become very rare. Next in time-scale comes the appliqué construction renowned for its embroidery (*tataura* – 'doing/making thread'), which consists of larger assemblages of floral blossom and leaves with embroidery to reflect shading. The flowers, stems and pollen are all replicated at least four times before being patched in a (usually) clockwise-rotational symmetric pattern on a large rectangular sheet of uniform colour. This kind of *tivaivai* may take weeks to complete, rather than several months, and is considered appropriate today as a wedding gift for a daughter or son. The simplest of all quilts is the cut-out or *manu*, which uses the well-known snowflake construction to arrive at a pattern that should have a well-balanced relation between background and foreground. Often a light material is sewn as a cut-out pattern onto a darker background, its fine lines giving the impression that the pattern resides in the darker spaces of the *tivaivai*. These *tivaivai* should take days to complete, although an experienced seamstress is needed to keep a straight and fine line that can be clearly traced on the back of the quilt. Such a *tivaivai* is appropriate as a gift to someone of the same generation or to a friend taking the place of brother or sister.

This preoccupation with valuing sewing in terms of time spent is a lasting reminder of the work of the London Missionary Society in the nineteenth century. The main Cook Islands Christian churches, whose buildings date back to the late nineteenth century, have numerous wall clocks (old and modern) decorating the areas around the pulpit and front entrance. Each appears to have been received as a gift from those who went to settle overseas, returning only for funerals and their own burial.

One could say that quilts and clocks travel in opposite directions, connecting many distant relations and linking households to one another across the Pacific and beyond. Yet the interrelation of clock and quilt movement in cycles of exchange is not only metaphoric, describing the relations that unfold over the course of a woman's life-time. We have to understand the pattern of the quilt as reconciling the *additive* nature of such exchanges with the clockwise rotational arrangement of cut-out motifs on the surface of the quilt. That Cook Islands patchwork patterns resemble the face of a mechanical clock and the rotating of its hands, is thus

The ultimate gift

In the beginning of my stay on Atiu I used to cringe whenever I saw *tivaivai* disappear into graves or being stepped on by a special person. 'What a lack of appreciation for all those countless hours of work and dedication,' I used to think; and in my 'disgusted, emancipated, European mind' associations like 'women's work being disregarded by being stepped on or buried and covered with dirt' would pop up.

The longer I lived on Atiu, the more I learnt that thoughts like these could only derive from the materialistic and self-centred Western world. Getting to know the people of the Cook Islands and trying to understand their culture, in many ways so very different to mine, this indignation has transformed into respect and appreciation for these people's ability to love and cherish their family. The appreciation of the work and time involved is not the chief concern here. It is the person who receives it that is. Precisely *because* of the knowledge, and appreciation, of how much time and skill it takes to make a *tivaivai*, there can be few better ways to express love and respect for a family member or friend. More important than the time and skill invested in the making of a *tivaivai* are the loving thoughts that are stitched into these coverlets, the purpose of which is ultimately to embrace the loved one for the final time and forever. What greater respect could one give to a deceased parent, partner or child than enveloping them in the most precious possession a woman has created?

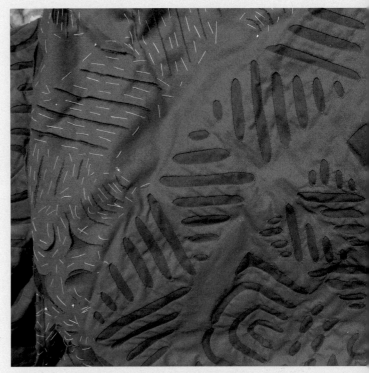

Tivaivai manu that Andrea is sewing for her husband (detail).

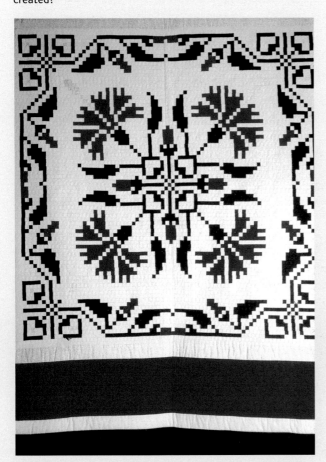

Tivaivai taorei that accompanied Andrea's father to his grave (detail).

I have learnt to see whether a *tivaivai* was made only to comply with the annual women's programme, or whether the seamstress spent the many hours dedicating every stitch to a person special to her. It is not just the bright colours and fancy stitches that make a *tivaivai* shine, it is the loving thoughts that make every detail precious. Most Cook Islanders know this and will not part with a *tivaivai* given to them, unless it is to show their own love and appreciation to another person special to them.

The first ever *tivaivai manu* I hand-sewed myself was made and given as a present to a dear Atiuan friend for his sixtieth birthday. A few years later, Papa Mariri passed away somewhat unexpectedly. As is the custom, his body was laid out on his bed, covered with layers and layers of *tivaivai*, for all of us to pay our last respects. When I stepped up to his bed to say goodbye, his wife who sat by his side pulled my sleeve. Silently, she lifted some *tivaivai* off the top to reveal the one I had made. I was deeply moved by this sign of acceptance. That was the moment I really understood.

I decided to sew a *tivaivai* as a gift to my parents on their fiftieth wedding anniversary. For ten years it decorated their bed. When my mother died, she was cremated and buried anonymously, as had been her last will. It was my father's wish to move to Atiu and live with us. We were fortunate to enjoy each other's company for three years. My father died in 2006, aged 94. He took this *tivaivai* to his grave.

Now I am sewing a new one. It was cut by a Marquesan friend and famous *tifaifai* designer with whom I exchanged a pattern. It has an incredibly elaborate design that will take me many years to finish. I am sewing it for my husband.. When I mentioned this to a friend, she remarked, marvelling at the pattern, 'You must love your husband very much and want him to become very old!'

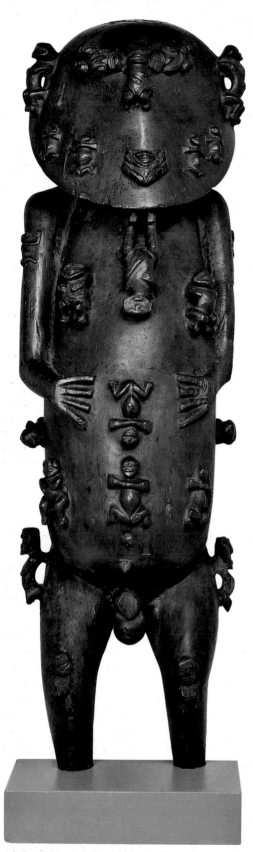

A wooden A'a god figure from Rurutu, Austral Islands. Height 117 cm. Acquired by the Revd John Williams in 1821. Ex collection London Missionary Society. British Museum, Oc, LMS 19.

not accidental, but an acknowledgement of the timely and incremental nature of sewing.

The marking of time: the quilt in a woman's life

Inspired by fibre-based techniques of lattice-work, the fabric quilts of the Pacific are key loci for knowledge about, and management of, socially effective biographical relations in ways which resonate with the modes of genealogical relationship and time, but which are still quite independent of kinship. *Tivaivai* making is of interest to people in the Cook Islands not because it sheds light on existing relations created through marriage or birth, which regulate the transmission of land and of entitlement to positions of influence and power, but because new relations of friendship, adoption and acquaintance can develop in the vicinity of a sewing. This thread, which binds people together who have no other relation with one another than the memory of a patched pattern, is the act of patching itself, which translates actual events in time into patterns that can be recalled over and over again.

Overtly, it is the biography of the maker that is traced in the stitched patterns of the large and elaborate appliqué and piecework quilt patterns (Rongokea 2001). Stored folded in treasure boxes, quilts in the Cooks and elsewhere in Eastern Polynesia are the tangible reminders of relationships that are thought integral to a person. It is symptomatic of this biographical aspect of patchwork that most women start sewing around the birth of their first child, as they begin to be active in the many exchanges that will connect their household to as many other households as possible during their life course. As every *tivaivai* is made to be gifted, and is often sewn with a specific occasion and recipient in mind, such relations are externalized into patterns that become iconic of events and relations. Patterns related to specific events may be repeated from memory later on in life, recalling the pattern in terms of the flower depicted, the symmetry used and, in the case of a piecework or *taorei* (composed of many thousand small pieces), in terms of numbers of coloured pieces of cloth joined to one another to make up the nucleus (*pu*) of the design that is repeated across the surface of the quilt. Patterns that have been gifted to, or buried with, a deceased relative are particularly strongly remembered, and may be re-made by a woman in anticipation of her own death, to be given to her surviving eldest child for use in her own burial.

This iconic potential of a quilt's pattern is shared by clothing, which is sewn for every occasion by Cook Islands women. Fabrics imported from Fiji come ready-printed with names of islands or popular events. Other dresses are made for occasions, such as the constitution festival or a church event, that demand a uniform dress code, with members of one family or one church wearing a dress or shirt of the same pattern and colour. The highlight of the dress-making competitions is the annual *tiare* or flower festival, when island dresses using imported and locally printed cloth are sewn and modelled by both young and old. Dresses that win prizes at such competitions rarely stay with the maker, as female relatives request them as tokens of their relationship.

It would be a mistake, however, to assume that a patchwork works merely as an analogue of the dress, which so clearly serves as a memento of an event or an emblem of a family relation. While it is certainly true that the particular type of stitch used in quilts, and even the manner of stitching, refers to the identity of the maker in ways that are clearly recognized by contemporaries in the competitive inter-island community, and while visually the pattern may be seen to serve as a memento of an event or a relationship specific to the biography of the maker (and which can thus be shared by relations who come into possession of the *tivaivai*), the pattern also transcends the personal space of biography by giving material expression to a de-centred vision of social relations, one which guides and directs the life-course of individuals through its binding logic.

There are qualitative distinctions between types of biographical relations that surface in the quilts' symmetries. There is the relation between grandparents and grandchildren, considered the most important relation of all and judged to be the backbone of a person as it provides both knowledge and entitlement to knowledge-based offices (both ritual and mundane). It is common for grandparents to adopt a child at birth and raise it as their own. There are relations that connect women of the same generation by virtue of being involved in the 'doing' of sewing; there are also site-specific relations, which women maintain through visits to other islands or in the wider diaspora. Each of these relations is expressed in distinct symmetries in *tivaivai*, in ways that bring out the transitive nature of such relations, which always entertain multiple points of view. What we see in the multiply repeated and non-random distribution of flowers on a quilt, is a kind of logic of affinity that allows one to see how women sustain relations among one another over time in certain predictable and idealized ways.

The fact that patchwork techniques and their specific symmetries are recognized as shared cultural knowledge across the Cooks and the diaspora indicates something very important that is highlighted by a comparison with the patchwork of colonial America, where collections document the almost infinite variety of quilting patterns in use among households headed by strong women, whose quilts are a legacy of their effective management of the home. In contrast to the emphasis on the individual emerging from the American record, in the Cooks the emphasis is on the 'socialness' of the memories that accrue on and around patchwork. Biographical relations in the Cook Islands are thus not unique to a single person, but instead take on an image of a social body.

The eradication of the distinction between singularity and plurality, between the one and the many, is forcefully visualized in the multiplication of a single motif whose symmetrical arrangement speaks eloquently about the relation that 'one' shares with the 'many'. This conversion of the singular into the plural and back again, made possible by distinctly fractal images, is one of the most distinctive cognitive functions of Cook Islands patchwork. What is striking about the fractal images on the surface of patched cloth is their striking structural and formal similarity to images from pre-colonial times, when men were in charge of matters relating to the political and ritual economy of the time.

Much has been written about 'fractal personhood' in the Pacific, where typical 'Western' oppositions between individual and society, parts and wholes, singular and plural do not apply, but where both persons and things and the relation between them is thought to be effected through acts of multiplication (Strathern 1988; Wagner 1992). No images have denoted these ideas of fractality more powerfully than the Cook Islands staff-god (see Chapter 1) and the Rurutuan treasure chest, in the shape of a figure symbolizing the god A'a. Notable examples of both are held in the British Museum, collected in the mid-nineteenth century during the early period of missionary influence in the Cook Islands. The anthropologist Alfred Gell (1998) has given us a most concise and interesting interpretation of such images, which he finds use scale to denote ways of thinking that are inherent in material forms. I give a short excerpt of his exposition to help illustrate the connection between museum artefacts and the *tivaivai* of today.

> What is particularly remarkable about the A'a is the explicit way in which this image of a 'singular' divinity represents divinity as an assemblage of relations, between (literally) homunculi. In so doing, the A'a obviates the contrast between one and many, and also between inner and outer. The surface of this image consists of amalgamated replications of itself, or alternatively, a succession of budding protuberances. Internally, the image consists of itself, replicated on a smaller scale, within its own interior cavity. As such, it images both the notion of personhood as the aggregate

He is I

To the German in me, coming from a meticulously regulated culture in which everything has its fixed place and purpose, its scientific explanation and its precisely descriptive name, the fluidity – the connectedness over and above the physical boundaries of time, space and identity – that I keep witnessing in the Cook Islands remains infinitely intriguing.

For the Atiuans of today, their brave ancestors come alive again in the legends and chants, whose recital is still part of a living tradition. I remember my fascination, when very early in our time on Atiu a neighbour stopped by our house on his way home from a *tumunu* (local 'bush beer' drinking session). He was in a happy, talkative mood, and soon began telling us the story of how his ancestors discovered Atiu. In the course of his story, his voice rose and his eyes sparkled, as he inadvertently slipped from the past tense to the present and 'he' – the hero – became 'I'. Alcohol had made it easier for him to gather the courage to tell us his story in English, but the way he told it revealed him as a true Polynesian at heart, whose native language knows no verbal forms to distinguish the past, present and future tenses, and whose sense of place in a long lineage of past and future embodiments of the same source made him assume, for an excited moment, his ancestor-hero's identity.

of external relations (the outcome of genealogy, fanning out in time and space) and at the same time the notion of personhood as the possession of an interior person, a homunculus, or, in this instance, an assemblage of homunculi. We cannot individuate the A'a in the way in which we normally individuate persons by identifying the boundaries of their person with the spatial boundaries of their bodies, for the A'a has no such boundaries, it is like a Russian doll, and in this respect, it irresistibly recalls the lines in *Peer Gynt* in which the hero compares the (moral, biographical) person to an onion, composed of a succession of concentric layers. (Gell 1998: 139).

Like the fractal image of personhood presented to us in the historic sculpture of the A'a, the Cook Islands *tivaivai* is an entity with relationships integrally implied in the ways motifs are serially enchained, 'budding' out of one another in a depiction of human life. Constructed from layers of additive pieces of fabric, cut and stitched into regular crystalline patterns on a plane, the Cook Islands *tivaivai* in fact presents the temporal equivalent of a spatial map in the form of a lattice of possible worlds. The idea of a person being composed of distinct layers of biographical relations impresses itself on the mind that visualizes the layering of the quilts that are wrapped around and folded on top of a person buried in the grave. We could thus think of patchwork as an old, and yet also new, means of gathering relations into a single fold, a move that has enabled Cook Islanders to sustain relations across time and space. Old in the way it binds together those that live apart, and new in the way it acknowledges that today effective relations incorporate not only relations based on birth or marriage, but also all the many circumstantial relations that life tends to fashion when it straddles the many places that mark the connections between diaspora and homeland.

Like all fibre arts made in Eastern Polynesian, *tivaivai* assume a central position as a supreme gift in exchanges in which they have, to a large extent, replaced both mats and barkcloth (cf. Weiner 1994; Hammond 1986). In the Cook Islands, *tivaivai* have come to act as substitute for both the cordage and barkcloth bound staff-god and barkcloth itself, which dominated the ritual economy during pre-colonial and pre-Christian times. The exchange of *tivaivai* in the Cook Islands of today is not only between persons or groups whose relations are built on existing reciprocity established through birth or marriage. Actually adoption (both formal and informal) is the most essential means of enlarging the radius of relations that may be encompassed in the social field that can be mapped by a woman's fabric creations during her lifetime. In women's gifting of sewing, *tivaivai* are fanned out in directions that seek to trace the paths taken by their children (both those they adopt and those that leave through adoption). The circuits of exchange within which a woman's quilts move during her lifetime touch and overlap relational fields that map temporal, biographical relations in a spatialized manner. Thus, despite the seemingly ego-centred production and circulation of *tivaivai*, their spatio-temporal distribution in fact resembles the mapping of space in the point-field mode of topographical modelling.

Women do not always keep track of where their quilts go. Though there are books that are kept to chart exchange transactions (similar to those that exist in many parts in Melanesia and, famously, in Japan), what is more important for them is keeping track of the patterns they have sewn over the years. Some keep boxes of tracings of cut-outs made for appliqué style *tivaivai*, and can tell long stories about what these would have looked like and when they would have been made, even if the sewing reaches decades into their past.

Contrary to the tendency to regard quilts as leftovers of a Pacific colonial past, we may thus instead come to understand their enduring popularity in terms of their ability to chart, visually and materially, a fractal vision of biography, and to make the charting of the path between possible past, present and future worlds comprehensible in ways that overcome the oppositions laid down by the homogenizing forces of globalization. Indeed, the re-examination of techniques of patching and its resulting symmetries as time-maps may allow us to understand why patching has emerged as one of the most successful metaphors of nationhood in the Pacific today.

Time-maps: quilt patterns from another point of view[2]

What is represented in a time-map, as Alfred Gell (1992: 242) reminded us, is not a unique or fixed state of affairs. Time-maps do not chart single worlds, but networks of possible worlds; for temporal cognition, like

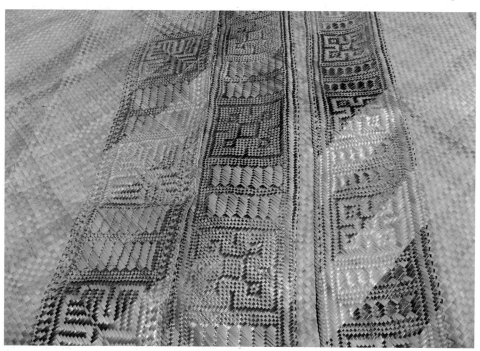

Cook Islands pandanus mats.

spatial cognition, has a logic by which it maps the paths that lead from one possible world to another. Like the stick charts of the Marshall Islanders, ritual artefacts in the Pacific call upon us to revise our conception of time and space within a surface/depth model that takes as its starting point the anthropomorphic figure of man (Kingston 2003). Rather than representing spatial relations from a particular ego-centred vantage point, Pacific representations of space and of time chart possible transformations in the continuous surface of two, quite distinct, media: that of cordage and that of lattice.

Elsewhere (Küchler 2003), I have argued for the pivotal importance of cordage as a medium for transformational imagery that is uniquely capable of mapping topological conceptions of space and time as links in a chain. The most famous example of the ritual significance of cordage is the Hawaiian sacred cord (*'aha*) which has also traditionally acted as a reference point for genealogy – representing not just the king's relationship with the gods, but also the connecting force that 'binds together all other genealogies, as it is the reference point of genealogical connectedness and the locus

of its legitimacy and truth', described by the anthropologist Valerio Valeri (1985) in his now classic *Kingship and Sacrifice in Hawaii*.

The binding of the cord as the core of the Hawaiian body politic resonates across the Pacific, and casts a shadow from the pre-colonial Pacific into the new world. From the ancient Tahitian knotted staff-god (*to'o*) to the Fijian spirit house (*bure kalou*) or the woven malanggan figure (*warwara*), cordage was key to transformational techniques of mapping, without being articulated everywhere in the same manner (Babadzan 1993; Küchler 2002; Rochette 2003). These examples of cordage-based ritual imagery point to its significance as a means of linking, through acts of binding and untying, that which is experienced as existing in the two separate temporal and spatial domains of the living and the dead.

The temporal logic of binding bore similar transformational capacity to spatial thinking in the pre-colonial Eastern Pacific. Cordage was a key material metaphor that resurfaced at key moments in the body politic when ideas of renewal came to the fore. In the Cooks a great number of binding

A handkerchief?

Every woman I asked told me that the translation of *taorei* was handkerchief. This sounded quite strange to me. How could a double-bed-sized coverlet constructed of thousands of one-inch (aprox. 2.5 cm) squares compare with a handkerchief? Handkerchiefs were not piecework, nor were they of the size to cover a double bed! So, why *taorei*?

My first clue was that the word did not so much refer to the finished cover as to the material from which they had once been made. The oldest pieced bedcovers that I had been shown, dating back to the early 1930s, were made of muslin, a medium-weight fabric that could easily be torn, rather than having to cut the strips with scissors. Reading the autobiography of the famous American anthropologist Margaret Mead supplied me with a lead in uncovering the mystery. During her first field trip in the 1920s, Margaret briefly stopped at Honolulu on her way to Samoa, the country in which she conducted the first studies that led her to world fame. A friend gave her a hundred small squares of ripped muslin 'to wipe the kids' noses' (Mead 1978: 118). The *tivaivai taorei* likely received its name from the basic materials it was made out of, coloured pieces of muslin, originally dedicated to serve as handkerchiefs, but 'misappropriated' by keen needlewomen to become the basis for intricate masterpieces of crafts(wo)manship. Life in places remote from supermarkets and department stores generates enormous creativity when it comes to using things for purposes other than those they were originally intended for.

I have yet another theory. Could *taorei* not be a transliteration of the English word 'dowry', the way it would be written phonetically by a Cook Islands Maori? Most women who had shown me their *tivaivai taorei* had told me they had received it from their mother as a wedding present. When collecting *tivaivai* for the annual exhibition at the National Museum, we usually ask the

owners to write down a few descriptive words in connection with the piece they have lent us (in order to give the viewers information regarding when it was made, by whom, whether for any special purpose, whom it belonged to, etc.) On one of the sheets I read: 'This presentation of gifts is referred to as an "Oora": mainly *tivaivai* mats, pillows, sheets, kapok mattresses and household gear etc. to help the young couple start a new life. This custom is rarely seen nowadays.' I was keen to consult my dictionaries, the older of which still refers to the tradition of using the word *'o'ora* for the ceremony to appease a war opponent, whereas in the more recent one the translation for *'o'ora* is 'dowry'. Even if the actual word can no longer be traced back to its true origin, my suspicion seems to have proved valid. The ceremony may have become rare on Rarotonga. Here on Atiu it is still frequent, at least in connection with weddings, though today it is called *akapareuanga*.

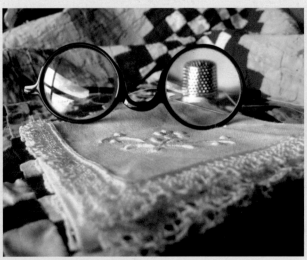

Why would a patchwork *tivaivai* be called 'handkerchief'?

Learning and teaching

At the Fibre Arts Studio, on Thursdays, the *mamas* would come and dedicate the day to working on their traditional crafts, which we would then sell in our Studio gallery. Since the plants that supplied the raw materials were quite exotic to me, I was naturally keen to learn about preparation and techniques.

The first task I was given was removing the spiky spines from the long, dried pandanus leaves, being careful not to pierce my fingers, and then rolling them up into bundles. Next I was shown how to smooth the leaves with the back of a knife, to make them more pliable. Eventually splitting the leaves lengthwise into even strips while leaving the butt end intact was also entrusted to me. Then the great moment came, when the *mamas* let me try plaiting for the first time: one up, one down, one over. All that had been easy, and I was quite proud of myself.

So much for the basic knowledge and tasks I was now given every Thursday. When I asked to be shown the more intricate parts of how to start and finish a mat, a basket or other item, however, suddenly the *mamas* would do things with such speed that my unobservant *papa'a* eye was unable to follow. It bothered me when my attempts of repeating what I had been shown turned out wrong and the *mamas* laughed. So I decided to take some pandanus home with me and try for as long as it would take to understand and get it right.

The next Thursday I proudly produced my work. The *mamas* obviously approved; even though I could not understand their Maori, I gathered that much from the look on their faces. From that time onwards, their teaching became slower, and they would be patient and repeat until they were sure I had understood. I learned how to prepare and dye the hibiscus bark that was overlaid on to the pandanus strips and plaited into the intricate patterns that adorned the borders of the more special mats. I learned how to make purses and baskets, hats and bowls. Soon I was even able to use my new knowledge to develop my own style and we would work out new items for our gallery together.

This experience made me reflect on the differences between my own Western culture and the Polynesian, until relatively recently oral, culture. In my school days, the students were

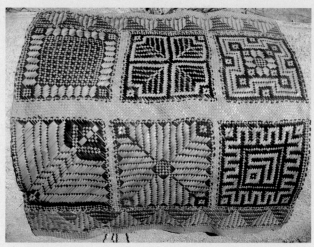

Sampler for mat patterns made at the Atiu Fibre Arts Studio *c.*1988.

supposed to learn certain things, interested or not, and teachers often talked to an audience that was not listening – the learning could always be done from a book close to exam time. In the Cooks, until the early nineteenth century, nothing could be recorded in writing. Someone's memory was the only container from which knowledge could be passed on to a future generation. It was therefore immensely important that the keeper of knowledge scrutinized very carefully those they would pass this knowledge on to. For one thing, knowledge is power and you do not want to empower someone who may misuse it. Secondly, it was important to ensure that the recipient of knowledge would comprehend and remember well, so they in turn could keep the knowledge safely. Having enough interest not to give in at the first failure, but instead to try until I had succeeded, qualified me for tuition.

Accidentally introduced pests have gradually infested nearly all the plants used for traditional crafts. Cheap imports of products from low-income countries have made it less and less interesting to produce one's own traditional items. The young generation have been more concerned with leaving the island to further their studies or to get well-paid jobs overseas. On our island, today, only the few surviving *mamas* in their seventies still know how to make a decent mat, hat, basket or bowl.

images existed, each owned by a family, a bloodline, a clan, a district, or even a whole island. By ordering these images according to size, the correlation between the ranked polity of images and social rank was given regular and formal expression in ritual.

The planar surface of containers plaited from plant fibre into latticework designs represented corporate clan unity as politically and economically situated in the present (as opposed to the temporal depth of kinship and immortality). While cordage can be extended, in principle infinitely, and requires a prior plan for proper execution, the shape and size of the surface of latticework depends on the constituent plant, such as the size of the leaf, either coconut or pandanus, from which a lattice-frame was created through a distinctive

bodily involvement with the product. The precise figuration of a latticework, whether a mat or a basket, thus differed with every product, while retaining its inherent reference to the plant that remains implicit within the materiality of its surface.

The contingent nature of lattice and cordage served as an outward sign of the specific connections that were believed to prevail among people in the pre-colonial Pacific. It was their enduring capacity to make visible an understanding of temporal connections that has given fibrous artefacts the status of an idealized model of sociality in today's world, which is now more dominated by the materiality of fabric and thread than by fibre itself. While cordage and lattice once carried opposing temporal maps of ancestral continuity and

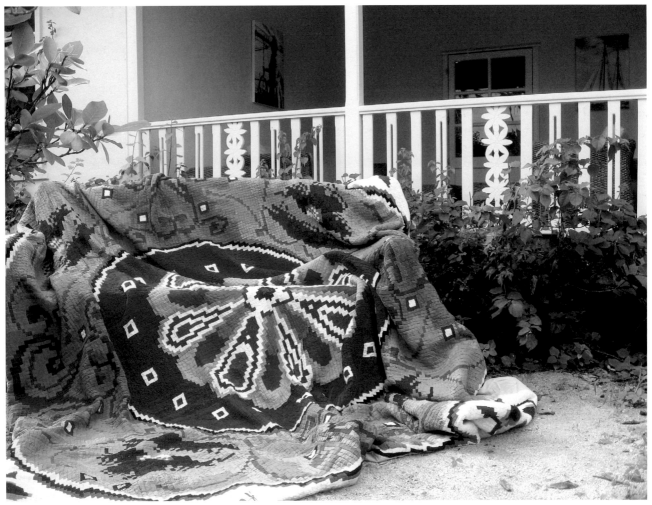

Tivaivai taorei by Vereara Maeva and her sewing group, from the collection of Rarotongan Beach Resort and Spa.

historical time, such an opposition no longer appears valid; for those who are living in the realm of the beyond, once occupied merely by the dead, are now the living, working for stretches at a time in the metropolises of the Pacific, only to return sporadically, and then finally to be buried in the homeland. Nothing could perhaps more poignantly underscore the integral part played by diasporic relations in the lives of those who stayed behind in the islands than the thread, which now stitches the fabric of biography into lattice-like patched cloth. Patchwork ingeniously combines the temporal worlds constructed through cordage and latticework into a single material frame that allows people to negotiate the complex nexus of relations that comprise transnational identity today.

Tivaivai have thus come to replace not just media such as cordage and barkcloth, used in the exchanges between polities, but also limited media such as mats that were used in the exchanges of the household. In the Cook Islands a so-called 'sleeping mat' (*moenga*) is still occasionally plaited from pandanus leaves, the pattern of the interlocking weaving being emphasized in the painted surface of the mat. The *moenga*'s pattern is used as inspiration for designs created by stitching together pieces of cloth, although graph paper, when available, is used to work out the number of coloured pieces that make up the distinctive pattern. *Moenga* are today rare in

the Cook Islands, both because of the difficulties of obtaining the raw material from the pandanus palm, and because of the ubiquity of *tivaivai* sewn from imported fibre, which have largely taken the place of mats in household exchanges. This displacement of the mat by fabric has led to a translation of the temporal mapping that is salient in the conception of the latticed mat into visually analogous images that are cut and stitched into the surface of fabric. The visual and conceptual relation between the mat and the fabric, and the temporal logic prevailing in both, is not unique to the Cook Islands, but is perhaps most clearly articulated there for reasons that may have to do with the legal implication of tracing connections in a regionally dislocated world dominated by a burgeoning diasporic movement.

As if to index the distinct biographical relations emerging from diverse contexts and directions of exchange, Cook Island *tivaivai* fall into three different types:

In *taorei* (*orei* = to wash) several thousand tiny square, diamond-shaped or hexagonal pieces of cloth are sewn together. The name possibly owes its origins to the second-hand cloth distributed as handkerchiefs by missionaries and reused for the patching of quilts in Polynesian households. Today, new cloth is bought in stores on the islands and graph paper is used to work out the design of the pattern, which

Tivaivai taorei – a challenge to your patience

Though *tivaivai* with hexagonal or diamond-shaped components are also sometimes referred to as *taorei*, symmetrically arranged mosaic patterns which are composed of fabric squares are most commonly associated with the term *tivaivai taorei*. *Taorei* can be translated as 'piecework', but there are also other kinds of piecing technique, such as *paka onu* and *etu*.

Taorei offers a uniquely Polynesian style of patterning and a variety of sewing techniques. Old mosaic piecework quilts were composed of patches as small as a square inch. For a decent sized bed-cover, over thirty thousand (!) pieces had to be joined, a task that needed the help of friends and family to get it finished in an acceptable period of time. Usually the pattern was divided into eight triangular segments, and was sewn by eight women. I can imagine that they would compete with one another to see who finished their part first, thus creating an incentive to get this massive job over and done with as quickly as possible.

Patterns of *tivaivai taorei* are nearly always symmetrical, though the symmetry does not result from individual patches having a symmetrical relationship to each other. Instead, they are arranged in blocks in such a manner that they form a mosaic image, which is then repeated symmetrically throughout the quilt: for instance, four flowers with corresponding leaves and buds rotated around a centre point. The centre may have an abstract, but symmetrically arranged pattern. The motifs, mostly flowers, are sometimes divided by abstractly patterned bands into a centre square and four corner triangles, or are dissected diagonally into four repeating triangles. In other *tivaivai taorei* we may find the motif (*pu*) repeated in a smaller version several times. There is no limit to the designer's creativity. Many versions of the same motif exist and show the distinctive style of a renowned *ta'unga* (expert).

Paka onu ('turtle's carapace', a wonderfully pictorial description) is a variety of piecework composed of hexagons. Unlike English piecing, in which the seamstress uses a paper foundation, in the Cook Islands *paka onu* hexagons are joined, and often even cut, freehand. It is amazing how regular in size they turn out, and with what admirable crafts(wo)manship they are joined. Both piecing and patterns are similar to (and probably derived from) American and English quilts.

Patterns are more often repetitive in *paka onu*. They are frequently designed in the round, and the motifs of some are quite abstract, requiring imagination from the viewer to 'see'. Some exquisite few have been worked in rows and show patterns inspired by nature, such as a starfish, or a picture that caught the fancy of the designer, like the popular anchor pattern derived from the label on the tins of milk powder that one can find in every Cook Islands household.

If the piecework components are diamonds forming star patterns, the local term is *etu* (star). Many years ago women copied foreign patchwork designs and arrived at a number of stunning works with a very Cook Islands touch. Instructions for hand- or machine-piecing these quilts have been published in many books.

The star shape results from joining great numbers of diamonds into striped triangles, which radiate and increase from a centre point, and then diminish again, forming 'peaks'. Nearly all examples I have seen show a large star in the centre, and sometimes smaller ones in the corners of the coverlet; however, the piecing is not continued to form an all-over pattern, but the stars are instead appliquéd to a backing fabric of contrasting colour.

Because they are so time-consuming to make, and require so much fabric, *tivaivai taorei* are amongst the more treasured *tivaivai*. It is therefore not surprising that the oldest *tivaivai* that can still be found in family glory boxes are nearly always *tivaivai taorei*. It is also the only *tivaivai* form that requires a backing, because of the need to protect the seam allowances.

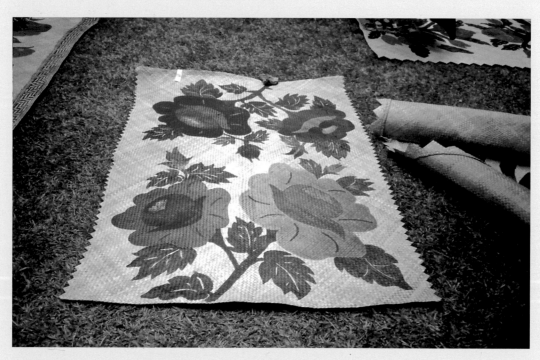

Pandanus mat painted with enamel paint.

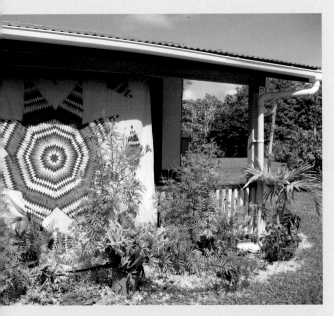

Tivaivai etu owned by Nga Mokoroa, Atiu.

I have read in some publications, and heard from some *mamas*, that in the very early days patterns were designed on mats because of the non-availability of graph paper. In fact, several annual women's groups' programmes on Atiu included a painted pandanus mat. The mats were plaited as is customary, and then a symmetrical *tivaivai*-style design was painted on top of them with enamel paint. I never saw a *tivaivai taorei* pattern derived from them, though. Some informants described their grandmothers or aunties as 'being able to put together a pattern from memory, visualizing the pattern as they called out the various pieces', certainly a great and admirable skill no longer around. I know the mother of a neighbour of mine had a pattern book from which she got her patterns. When I had a look at it, it seemed to be of Danish origin – or at least in a Scandinavian language – and was an instruction book for counted cross-stitch. Unfortunately, this was many years ago and the book has since been lost.

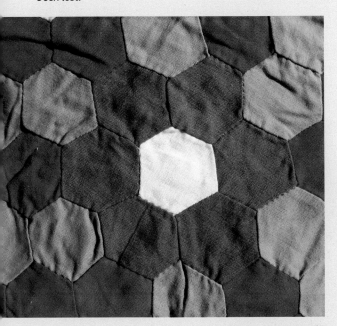

Typical *tivaivai paka onu* (detail).

can be given varying degrees of complexity without losing its essential logic of composition.

Taorei always consist of a single, multiply replicated motif (*pu*) connected by trails (*tarere*), creating an asymmetrical chequerboard pattern. There are two principal variations. The construction of this *tivaivai* can involve the fabrication of blocks that are repeated across the surface of the fabric in a reflexive, transitive and yet asymmetric manner; that is, identical blocks are linked so that the eye can pass from one to the other, often without the emerging pattern being reducible to a point symmetry (see page 105). By repeating the core motif in the round, rather than along the horizontal plane of the sewing, a new motif may appear which consists of enchained and integrally related fractal components that are usually repeated in eight triangles from a central point. The basic shape used for this construction is a square. Central to the construction of this pattern are mathematical conceptions, as tiny coloured pieces of shredded cloth are threaded in exact number and sequence onto a string before being re-stitched into a pattern. Ideally, the fractal pieces of the overall pattern are to be sewn by different women, usually by sisters or, today, work colleagues, though in practice women often find it conducive to the uniformity of the sewing to carry the work out themselves.

Individual pieces of the *taorei* can be kept and used as a template for recreating the pattern when the sewing has begun to disintegrate. *Taorei* are personal possessions that, if they are not wrapped around the owner at death, are gifted as a token of 'vertical' and asymmetrical succession. As tokens of succession, such quilts hardly ever leave the household and tend to be retained in the original owner's treasure box.

The most popular sewing, which circulates most rapidly between related households in successive gifting occasions, is called *tataura*, and is characterized by appliqué and embroidery, usually of floral motifs with leaves. The use of multicoloured thread in the overstitching of embroidery gives the design the impression of being three-dimensional; care is also taken to underscore the repetition of a singular motif by repeating the embroidery on each appliqué piece with precision. Distinctive to the *tataura* is the manifold replication of identical *pu*, motifs that usually consist of a single flower with its leaves and blossoms, which are positioned on the surface of the fabric with a rotating or diagonally offset symmetry that is both reflexive and transitive, in that the pairing of two motifs allows the remaining composition to be logically deduced. Such *tataura* are the preferred gift at weddings and sons' hair-cutting ceremonies, when they inspire recollections of 'horizontal', that is temporally coexisting, genealogical connections formed through marriage, which link islands, villages and named clans across generations.

The third type of technique is called *manu* (bird), a cut-out that is superimposed upon a background of opposing colour. In construction, the material is folded twice (in half and in quarter) to create four segments when unfolded, or three times (in half, in quarter and diagonally) to create eight segments, which unfold after cutting the design into it. Care is taken to create an even balance of positive and negative space, with the motif unfolding along the central

Dances on fabric

By far the most stunning works are the embroidered appliqué covers. They display Cook Islands women's strength and their love for colour and drama.

As a trained embroiderer, this was the technique I most wanted to learn. My first self-designed *tivaivai tataura* was far from successful. Technically, it may have been a good work, but as I learnt more about Cook Islands women and their ways I realized that technical perfection played only a minor part in achieving the desired effect. It was perhaps this amazing species of textile art that told me most about the women in my neighbourhood.

Women on Atiu are not tiny and quiet, and do not tuck in their bellies or their feelings as European women were taught to. They are big and bold and noisy and all-embracing. They are the strength behind a family; the majority are participants in public meetings, not afraid of speaking up and debating decisions, and they are the ones who really make things happen in this country. Meanwhile the men, many of them known to be great orators, talk.

Art in the Cook Islands is often understood as performing arts. People here have music and rhythm in their veins and are trained to appear on stage from an early age. *Tivaivai tataura* are public performances, they have music (colour) and rhythm (composition), and their impact when seen from far away causes a feeling in your belly much like the vibrations of drumming.

Like *tivaivai taorei*, work on the embroidered covers can also be shared with other women and is therefore often a group activity. Once the pattern parts are cut out, the design lines are traced, and the various stitches and threads are decided upon, the first segment's parts are embroidered and act as a prototype for the other helpers to copy. Stitches are not necessarily discussed with words: *tivaivai* making is a visual activity. This made it particularly hard for me to discover commonly used names for embroidery stitches. Many

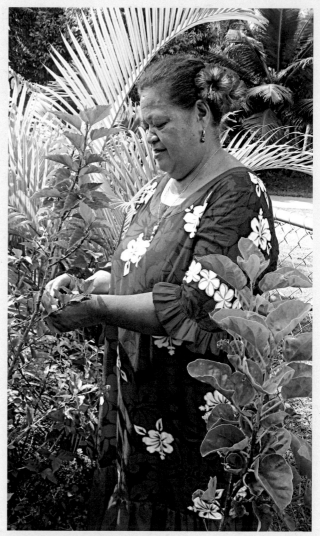

Cook Islands women are big and all-embracing.

women I asked during my research knew neither English nor Maori names, though they could show me how to sew the stitches.

Unlike Western and Asian art where often 'less is more', *tivaivai* are at their best when displaying sewing and embroidery skills to their maximum. Cook Islands designers of *tivaivai tataura* take great pride in demonstrating how many stitches they know. Some even invent their own varieties, giving their works a unique personal touch.

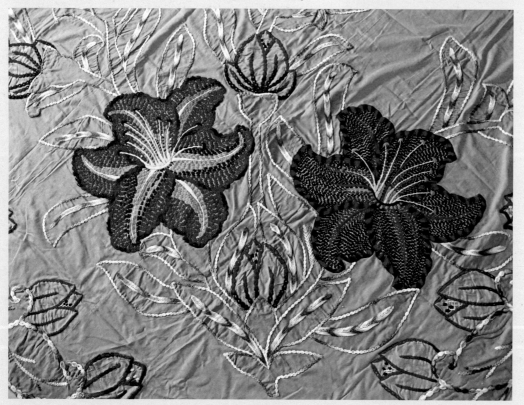

Tivaivai tataura embroidered by Parau Taruia (detail).

axes in symmetrical fashion. Both the cutting of the material and the sewing have to show a clear line, visible on the back of the fabric, that moves without stopping, with no obvious beginning or end. Tracings of cut-out designs are sometimes drawn on tissue paper or clear plastic and carefully stored for future reference. Such *manu* are the quickest *tivaivai* to make and, as they generally do not involve more than two colours, are less expensive to produce, or arduous to plan, than other types of sewing. They are the preferred gift for those outside the immediate family, are readily 'lent' to friends who need help in acquiring sewing that can be gifted, or are even sold to strangers. Often lost in exchanges, the pattern of this *tivaivai* can be remade repeatedly.

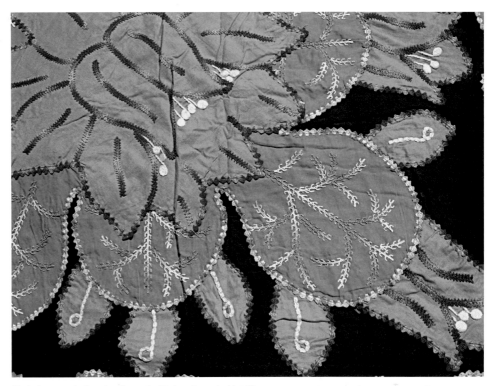

Tivaivai tataura designed and sewn by Kimiora Samuela (detail).

Such is the presence of sewing in the life of women in the islands and diaspora communities of New Zealand and Australia alike, that the stability of the three generic types of techniques is surprising. If the Cook Islands patchwork styles were merely the result of inter-island borrowings of ideas and consisted solely of the modification of artistic principles in line with regional aesthetics, as Joyce Hammond (1986) has suggested, one would expect a much greater variety of techniques and visually distinctive patterns, given the mobility of Cook Islanders. In fact, Cook Islands patterns merely differ from those produced in other parts of the Pacific in the choice of motifs and of colour combinations, yet they consistently deploy the same manner of connecting motifs on the planar surface of the fabric as other Polynesian communities. Here and elsewhere, the attachment women have to the *tivaivai* they have sewn is intense, suggesting that the ideas invested in the materials of cloth and thread are profound and capable of charging new relations.

As Elizabeth Akana (1986: 5) says, about Hawaiian quilts, 'quilts contain much *mana*. If one should die and leave much spirit behind, it could be damaging and might never be able to achieve real rest.' The intense attachment of person and quilt means that quilts are only separated temporarily from the maker, eventually being returned at death. While Hawaiian quilts were burned, Cook Islands *tivaivai* are buried with the dead in concrete, tomb-like structures erected next to the entrance of houses.

As quilt-like sewings are in effect the burial shrouds of their maker, anticipating a future state of being where all possible past and present temporal worlds coalesce, a deep association – which traverses the production, the circulation and the 'style' of the pattern – exists between biographical time and patching. *Tivaivai* in fact resonate with temporal logic in generalized ways that go beyond individual recollections of biographical time to the non-causal, but *modalized*, that is logically describable, enchainment of events in what Alfred Gell called 'B-series time' (Gell 1992: 238–41).

B-series time, according to Gell, reflects the temporal relationships between events as they really are, out there, while the A-series provides subjective and tensed perceptions of events that happen in the B-series temporal territory. In order to know how to act in a 'timely' manner, to plan and anticipate what might happen, but also to defend claims about the past, we have to construct representations of B-series time, so called time-maps, that allow us to 'navigate' in time based on an understanding of how possible worlds are connected with one another (Gell ibid.: 235–6). Time-maps, like maps we use to navigate in space, rely on the generation of images that correspond to how the world appears to us; these images would be useless for navigation, however, unless they are visualized in a map-like manner. This translation of non-indexical space-time beliefs into images, and then back into maps, creates what Gell called 'indexical fixes' or perceptual beliefs which give rise to our inward sense of time (Gell ibid.: 236–7). Image-based maps of B-series time are not the 'real thing', but as no experience is possible of 'real' four-dimensional space-time, we are forced to rely on such reconstructions in mastering time.

The mastering of time as genealogical knowledge is of fundamental importance in the Cook Islands, where space is at a premium and where locating oneself in a time-map of social relations is the prerequisite for being granted right to land. When sometimes more than 48 generations can be recounted as coming together in one person, a fractal map that depicts the enchainment of social relations serves as a model of relationality that can be used to make decisions whenever matters of attachment are at stake.

Tivaivai as time-maps thus evoke a world of images that are derived from subject-centred time awareness (in that they are in and of a specific place and time), and yet also match these images against invariant templates derived from the underlying cognitive maps of the B-series (Gell ibid: 254). As maps, image-based representations such as the tivaivai of the Cook Islands draw out as temporal beliefs the logic of internal, template-like representations of time. Tivaivai seen as time-maps are thus ordered in highly standardized and systemic manners as they depict a modal logic that tells us about all the possibilities that are deemed feasible to connect events, so that one may think about them as before or after, necessary or possible.

The techniques of connecting specific to each of the three Cook Islands tivaivai displays precisely such a modal time-map: they do not reflect subjective and tensed perceptions of events, or culturally specific beliefs of how events in past, present and future are disposed towards one another, but a generalized temporal logic; and it is because the mapping of time is logical, while reflecting specific local beliefs and attitudes to biographical time, that Cook-Islands-style tivaivai are recognizable and yet distinctive across Eastern Polynesia and beyond.

The linear arrangement of the cut-out (manu) recalls the linear enchainment of possible worlds in which specific paths are singled out that connect past, present and future. The cognitive activity of 'projecting' that is involved in giving time-maps linearity in this manner is fundamental to the production of the cut-out, which, when folded, requires one to envision the cut lines projected symmetrically across the missing planes in order to arrive at a recognizable image of a flower or plant when the fabric is unfolded.

In contrast to the cut-out, no unique past, present and future is represented in the embroidered appliqué (tataura): here repeated, identical images, usually of flowers, are arranged, mostly in rotational or diagonally offset symmetry, to depict the reflexive, transitive, but symmetrical relation between possible and co-existing worlds. By visualising equivalences between images, a singular, repeated motif is arranged in a precise symmetrical manner on the tivaivai, and no significant relational structure emerges.

In distinction to the relational symmetry of the appliqué, piecework (taorei) images are arranged in an asymmetric manner. Interconnecting paths (vava'i – to split to break up, to divide) mirror the before and after relation between successive images (pu) of possible worlds that, as in the appliqué quilt, are identical to each other. Past, present and future are depicted as relational, but not unique, and as reflexive, transitive and progressive.

Rather than assuming that women project their personal biographies into the quilts they are stitching, the tivaivai as time-map enables women to project an understanding of their lives in ways that speaks to others. One could even go further and say that it is because tivaivai comprehend time by locating subjective images, sometimes recalling specific datable events, in a logical and non-indexical map, that they can figure as the quintessential inalienable exchange object that enables one to 'keep while giving' (Weiner 1992). We can now begin to imagine how patching may enable women to manage time, and relations unfolding in time, in ways that are made effective in the performance of exchange, when patterns not only connect persons but link past, present and future in ways that resonate with temporal beliefs.

As they are agents for the navigating of time, tivaivai have real consequences for people's lives; their functions are far weightier than mere 'decoration'. Tivaivai are implements for the management of temporal beliefs that are perceived as vital to health, wealth, progeny and salvation. It is thus also for ultimately practical reasons that time-maps are logically conceived, rather than being merely richly symbolic or indexical.

It would be stating merely the obvious to say that every tivaivai is valued with respect to the time it took to be sewn. Much of what can be said

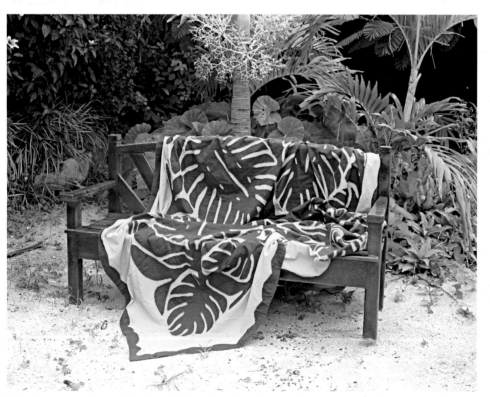

Tivaivai manu showing leaves of the 'Fruit Salad Plant', from the collection of the Rarotongan Beach Resort and Spa.

Migrating birds?

All my informants told me that the translation for *manu* is bird, insect, animal or any living thing moving on the earth (Savage 1980: 139). It struck me as strange, though, that in almost all cases the designs displayed on *tivaivai manu* are flowers. Hardly ever, with the exception of butterflies, can one find birds, insects, animals or human forms used as motifs on old *tivaivai manu*. The true reason for this term being used to describe a two-layered coverlet whose top layer has been folded and cut-out 'snowflake fashion' is no longer known.

In my research I came across two explanations, both referring back to the time when ceremonial cloths were made from tapa.

Kite flying was a popular pastime in the pre-missionary Cook Islands. The kites, which could have enormous extensions, would be covered with a special tapa cloth made from paper mulberry bark. This tapa was called *tapa manu, manu* in this context being the word for kite. Freehand drawings would often decorate these giant 'birds', so consequently *tapa manu* was a cloth with decoration drawn freehand.

The Samoan word *mamanu* described something similar. In Samoa, women had two kinds of decorated tapa: *siapo* (barkcloth) *tasina,* which has been decorated by spreading over a tablet carved in bas-relief and rubbing with vegetable dyes; and *siapo mamanu*, which has decoration drawn freehand.

The time of the introduction of *tifaifai* (the Tahitian word; as presumably they were introduced to Tahitians first) was when missionaries were travelling back and forth between the Polynesian countries to exchange their experiences; traders and whalers were buying and no doubt exchanged 'curios'. In this context of cultural transmission, perhaps the Samoan *siapo mamanu* helped to baptize the snowflake-style *tivaivai* designs of the Cook Islands with the name *tivaivai manu*.

The technique

Technically, *tivaivai manu* can best be described as snowflake-style cut-out appliqué. Paper cutting is an ancient art form that is said to have originated in China thousands of years ago. Whether the artistry of Tahiti-based Chinese immigrants influenced the creation of this patterning technique and its application to fabric, or whether the French Polynesian islanders developed this cut-out technique some other way, is unknown.

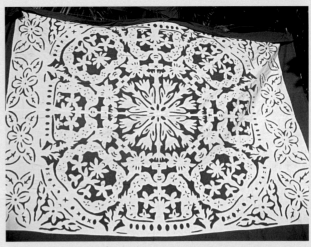

Tivaivai manu 'Crowns' cut by Patikura Jim for Tira Tararo.

Most probably, all of us have folded a piece of paper, cut shapes out of it, and unfolded it again to marvel at the result. What is the secret of this marvellous transformation of a simple shape into a wonderful pattern?

Symmetry.

Some call the result of our folding and cutting exercise 'paper doilies', others 'snowflakes'; in my native country it is called 'Scherenschnitt', a name based on the technique used (cutting with scissors) rather than the finished result. By any name, this technique is the way you design a *tivaivai manu*: by creating with fabric and scissors shapes that are arranged repetitively on either side of dividing lines or axes.

The Hawaiians fold their fabrics into eight segments to obtain intricate pattern repetition; however, the fabrics are much thinner than those used in the Cooks, because their quilts are padded and lined to give them body. In Tahiti you see many two-fold designs (four segments), as Tahitian artists use commercial bed-sheets for making *tifaifai*. One is cut into a pattern, leaving the edges intact so that they can be joined to the edges of the second sheet, which serves as the background. In the Cook Islands many patterns are square and are just at the centre, or they have a centre design and a different design band on either (or both) of the short sides (if the cover is rectangular). Therefore, you find both an eight-segmented and a four-segmented symmetry here, even though four layers of heavier fabric (such as broadcloth, here called Azlin or homespun) are easier to cut than eight.

about the temporal logic underlying them could be extended to mats, but the visual and material differentiation facilitated by the sewing techniques of *tivaivai* amplifies some of the more salient practical reasoning. Sewing has enabled time-maps to be shared and modalized, describing matters to which we have no access in human experience in ways that allow temporal worlds to be charted predictably.

By mapping the otherwise inaccessible territory of four-dimensional space-time (once thought to be the ancestral sphere), patchwork took over where ritual stopped. With the introduction of Christianity, and the demise of ritual, cordage-based imagery has given way to new thread-based techniques of image making, such as sewing and embroidery. Just as the touching of the 'stick-god' was restricted to the momentary ritual 'undressing' of the gods at the height of pre-Christian ritual so brilliantly described by Babadzan (1993) for ancient Tahiti, so sewing today is taken out of its storage box only once a year, to be shaken out, cleaned, inspected and, if need be, mended by being given a new backing that covers it when it is folded up again and returned to the box until its next revelation (like the dressing of the staff-god with a new layer

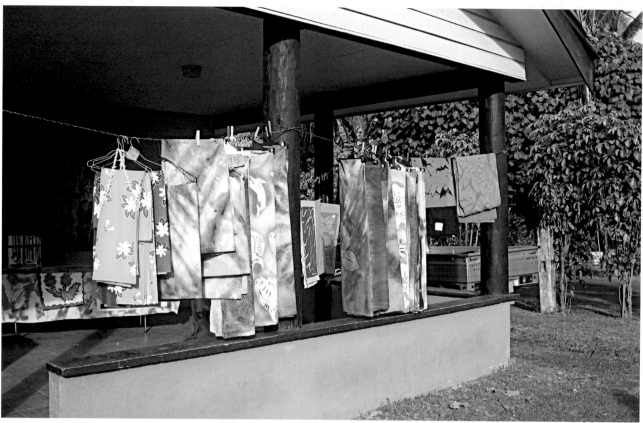

Dyed *pareu* and tacked *tivaivai* offered for sale at a road-side stall on Rarotonga.

of cloth). Genealogies, or biographically effective relations, are similarly recited whenever *tivaivai* are gifted.

With this new-found appeal of thread-based images, the once quite distinct logic of mapping evident in latticed things became blurred with that of cordage in the fashioning of patched and embroidered quilt-like sewings. What emerged in the conflation of cordage and lattice-based techniques of mapping in the sewing patchwork, are unique time-maps that elicit a temporal logic in which ideas of continuity through acts of renewal merged with empowerment through acts of containment. Patchwork thus gave rise to a political economy that is centred on biographical relations. Spatial and temporal cognition are key factors in the consolidation of effective biographical relations, which are known and remembered through sewing, the material connectivity of which asserts that relations are made and not born.

We have seen in this chapter how patchwork fuels an economy run on 'past'-time in which generalized commodity value is converted into an asset that sustains the economy of the household. Women who sew together regularly share assets such as bank accounts into which they pay proceeds of occasional sales of *tivaivai* and from which they draw money to purchase the cloth required for a new sewing. Sewing bees such as these, however, also sustain the household as whole, as they regularly inspect each others' homes, and purchase from the collective account what is needed to keep them running. What may appear to the innocent eye as an ostentatious craft made by the busy hands of quiet women thus challenges our expectations with an astonishing testimony to the pooling of assets in things and people that is vibrant and sustainable in the homelands and elsewhere.

Notes

1 Lehman and Herdrich point out the importance of the distinction between the two modes of representing space in their linguistic and ethnographic research in Thailand and Samoa (2002: 179–97).

2 A version of the ideas expressed over the following pages has previously been published in *RES* (Küchler 2005).

Tutaka: humble homes become colourful castles

In *Ancient Tahiti* (1928: 158–60) Teuira Henry reports that the priest used to change the god's tapa and feather decoration every three months. Only priests were entitled to plant the special *aute* (paper mulberry – *Broussonetia papyrifera*), to prepare the cloths to cover the *to'o* (family god) and to deal with the *tapu* of the gods, as only they had the requisite secret incantations with which to protect themselves. The *marae*'s treasures were kept in an *ia manaha* (sacred storehouse) and had to be aired every three to four months to prevent them from falling prey to the mould and destructive bugs that abound in warm, humid climates.

In 1926 the *Au Vaine* (lit. 'the women') established the *tutaka* ('inspection'). The heathen gods had long since been abolished and all Cook Islanders had been converted to Christianity. But the custom of cleaning and airing all cloths and textiles regularly, which in Henry's book was connected to priests and sacred dwellings, seems to have remained and has become important for people's own homes. This three-monthly inspection tour ensures that every household benefits from regular efforts to keep it clean and its inhabitants healthy.

If you visit Atiu during the days preceding the *tutaka*, the uniform buzz of lawnmowers and grass-cutters will greet you as you enter a village. You will see immaculately manicured gardens with long washing-lines decorated with colourful clothes next to every household, and you may even find all mats, cushions and mattresses spread out on the lawns around the houses to air.

I would like to share with you some early memories of our life on Atiu. When I first learned what *tutaka* was, I shuddered! Atiu, in those days, had only dirt roads. And some dirt it was… The oily rust-coloured earth entered the louvred windows in clouds whenever a vehicle drove past. Admittedly there were

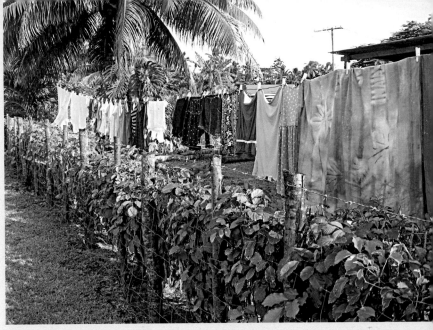

Colourful clothes-lines adorn every household before *tutaka* time.

fewer vehicles then, but enough to have kept you busy all day just wiping dust, had you intended to keep your place clean and spotless at all times.

The *tutaka* committee included the doctor, public nurse and health inspector, and altogether was a group of approximately ten people. These would not only want to examine your dust-free, sparkling home, but would also look under every bed, inside every cupboard and at every shelf, your suitcases (local substitute for wardrobes) and camphor boxes filled with neatly folded, recently washed and aired laundry. They would check cooking-house, wash-house and outhouse, and judge the results of your gardening skills and your taste in decorating! I did my best, I scrubbed, washed (with the next-door neighbours' girl's help) and polished, and got Jürgen, Mr Green Thumbs, to plant two frangipani trees and a bougainvillea.

When the big day came, I waited and – saw the committee go past without even a look at the results of my hard work! Frustrated, I ran after them and dragged them to my house: what else had I worked so hard for, if not for them to come and check and approve what they saw! After the inspectors' shyness had passed and their 'ohs' and 'ahs' had stopped, I was told that *papaā* (European) houses did not have to be inspected, because *papaā* were always clean and orderly (little did they know, but I wasn't going to tell them!)

Upon my further inquiries regarding the custom of *tutaka*, I learnt that the next one, at the end of the year, would be a special inspection for which women prepare all year long. The main emphasis in that *tutaka* is on the house's decoration: covers for sofas, chairs and cushions, tablecloths, bedspreads and pillow cases; even floor coverings and doormats. After finishing the inspection, there would be a grand ball in the Atiu Nui Hall and a trophy would be presented to the winning village. In good Polynesian custom, the winner is not an individual, but the village community as a whole. The village whose women have all fulfilled the demands of the women's committee's programme, decided upon at the beginning of the year, will win. Often there are several winners and the cup is shared, like so many other things in our island community.

The *tutaka* is currently practised with three-monthly regularity on most islands of the Cooks group.

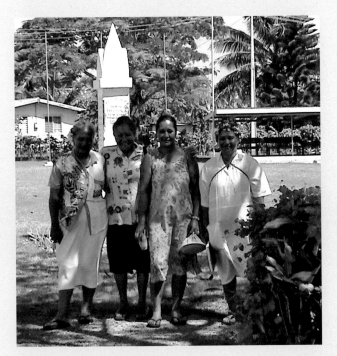

The *tutaka* committee on its way to work.

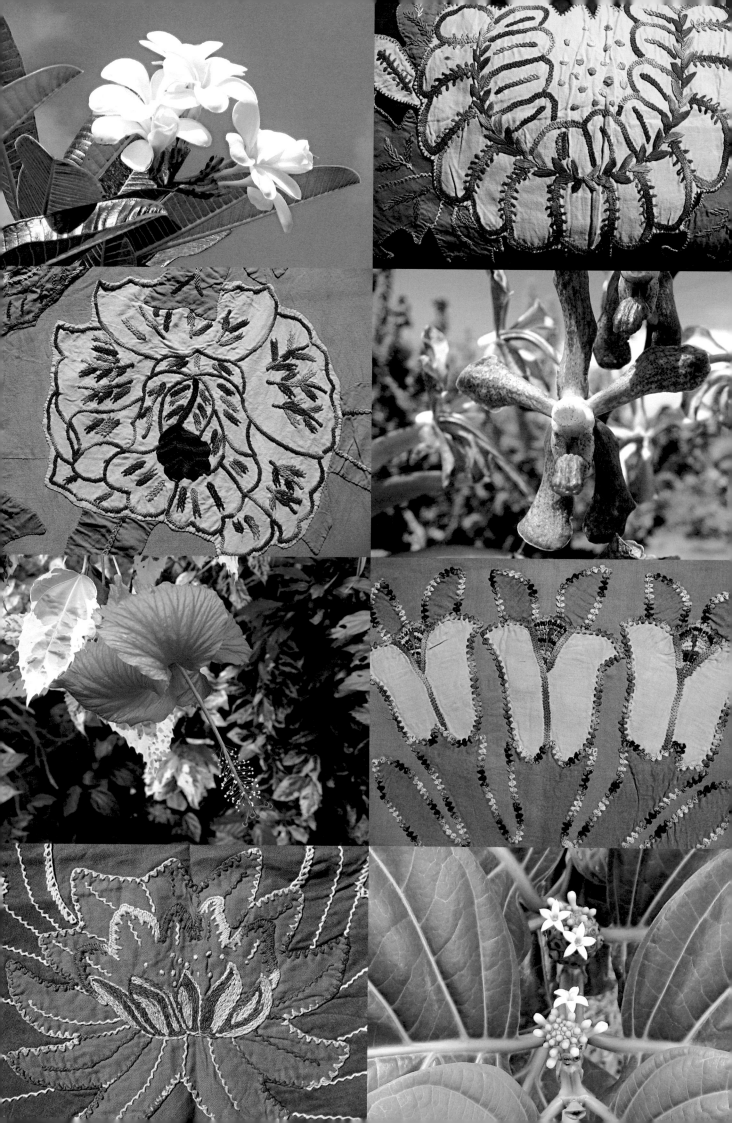

Chapter 4

Hidden in the light

Characteristic of Cook Islands *tivaivai* is the ostentatious depiction of flowers, arranged on large, coloured sheets of cloth. Each *tivaivai* depicts one or more flowers repeated many times over its surface in a symmetrical pattern. The flowers are shown most vividly, and with striking verisimilitude, on appliqué *tivaivai*; while piecework and snowflake designs occasionally obfuscate botanical identities by means of geometric abstraction.

As one surveys the plant kingdom modelled in *tivaivai*, one is reminded of a botanical manual of archetypal images. Yet one also quickly notices with astonishment that many of the flowers that are depicted with such attention to detail are strangely familiar, bringing to mind gardens from temperate climates, such as those found in England or North America. There are the beloved roses, marguerites (the English daisy,

not to be confused with the Cook Islands word *matirita*, which refers to the English chrysanthemum), gardenias and pansies; but one will search in vain for such Cook Islands icons as the frangipani flower, the *nono* plant (*Morinda citrifolia*) or even the hibiscus, found all over the islands, but rarely finding its way onto *tivaivai*. It is the foreign, the strange and the forgotten which appear to be collected up and stitched into the fabric, where they are held in place in the form of memorable images. One can discern this trend from pre-Christian times in the binding of the feathers of foreign birds (which served as tokens of the omniscient and yet distant power of the gods) into cordage-wrapped wooden staff-gods by means of floral-shaped knotted cords, or into male ritual paraphernalia such as feather headdresses or sculptures.

Flowers, natural and sewn (clockwise from top left): white frangipani, chrysanthemum, orchids, tulips, *nono*, waterlily, hibiscus, rose.

Pansies

In 1991, after an article about our company was published (Eimke 1989), the Atiu Fibre Arts Studio was contacted by the Folk Art Museum in Fort Mason (San Francisco, USA), which was interested in exhibiting Cook Islands *tivaivai*. In early 1992 we packed a good cross section into two metal boxes and shipped them to San Francisco. Mata Teamoke, then my co-director, and I flew over to attend the exhibition opening.

The museum was quite tiny, but the curator had done an exquisite job in displaying each work to its best advantage. A television crew came for interviews, and we were quite proud to see our island's treasures so well respected.

During this week in April we were lucky with the weather. It was possible to walk to the museum from our accommodation, and in these daily journeys we passed many houses with window-boxes of colourful flowers. Whenever we passed those that displayed pansies, which had so many different sizes and colours, Mata would stop me and have a close look. She could not get enough of them. Pansies cannot grow on Atiu, as it is too hot, but I had the feeling that that was not the reason that Mata was marvelling at their beauty and variety. 'No!' she laughed. 'We are planning on a programme [two pillow cases, a table cover, a sofa cover and two cushion covers!] for our *tivaivai* show which will all

be "pansies" *tivaivai tataura* having three colours in the flower and green leaves.'

That year's show was one whose wealth of colours and number of participants (400) remain unrivalled. Needless to say, Mata's pansies had the most realistic colour combinations and shapes...

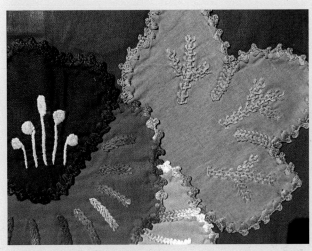
Embroidery (detail) showing stylized pansies.

The passion-fruit vine

The entire family came together to celebrate the unveiling of Papa Sam's headstone in January 2008, and during the event one of his sons-in-law read a story, 'My one-legged Papa Sam', written by his grandson. Despite the solemn occasion, we all laughed, as that's how many remember Papa Sam in old age: sitting in the garden, confined to his wheelchair, but full of humour, just as the boy's story reminded us.

In his younger days, Papa Sam had been a keen rugby player and a skilled carpenter. He had built houses, and had adorned the precious native hardwood beams of the village meeting house with expertly carved patterns. Later, all the children came to know him as Enuamanu School's caretaker.

Badly healed rugby injuries and diabetes ultimately cost him one leg. At the age of 75, he felt too old to learn to use a prosthesis, and instead opted for the wheelchair that his children and grandchildren pushed through the streets of the village or into the garden, where he preferred to sit.

His favourite spot was in the shade of a passion-fruit vine. There is too much shade for the plant to bear fruit, but his daughter Tangata keeps it growing and has defended it from family members wanting to cut down the 'useless' plant. 'It reminds me of him', she says.

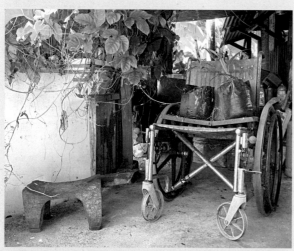

Papa Sam's favourite place under the passion-fruit vine.

While taking photographs during the unveiling ceremony, I discovered the wheelchair, now supporting new plants in their planter bags, and with potted ones around it – all waiting to be placed by his grave once the family members have returned to their homes on other islands and in New Zealand and Australia. 'I was going to put his wheelchair next to his grave,' Tangata says with a smile, 'because that's how we remember him! He loved his garden. When my sister brought out the photographs now included in our family genealogy book, we noticed that this garden is the background to almost every photograph.' Like Papa Sam's passion-fruit vine, connecting the branches of the tree in which it grows, so does this garden grow through the memories of a family branched out over oceans, protecting the soil in which many of them will ultimately rest.

When we enquire about the origins of images that ostensibly remind the viewer of distant landscapes so very different to the rocky outcrops of the southern Cooks, we are directed to look at the garden, which is to be found just outside the front porch of every house. Here, potted plants are grown that are the pride and joy of those living inside. Every plant has a story, a recollection of how it came to the island as a cutting, brought back from one of the many trips people make to relatives living abroad, and of how it was passed on as a seedling among relations on the island. These potted plants are sometimes placed on or around the roofed grave structure that tends to be built next to the veranda on which women sit and stitch their *tivaivai*, and which are a central conversation topic of every visit. Potted, the plant is ready to be depicted, described, interpreted and transported. Tended in this manner, away from their natural setting, the potted plants and their depictions do much more than show nature 'only', but rather display visions of self and society that we should understand as crucial and revealing constructs.

Less openly declared referents of stitched flowers are to be found in women's strictly guarded scrapbooks, which contain tracings, drawings and cuttings from embroidery and sewing magazines that have somehow found their way to the Cooks. Many of these scrapbooks are thought to have been lost to time, a loss that is compared to the much feared loss of the gardening knowledge that allows foreign plants to thrive in new places. Others, usually *ta'unga* or their descendants, have boxes, carefully stored under the bed, that are full of paper cut-outs for appliqué and snowflake *tivaivai*, as well as

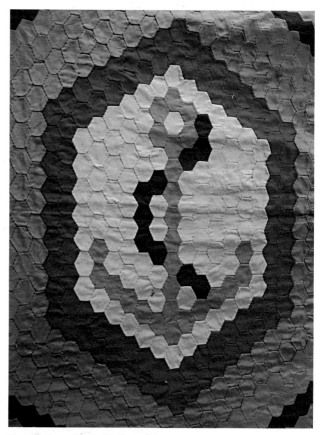

Tivaivai paka onu 'Anchor' by Vereara Maeva.

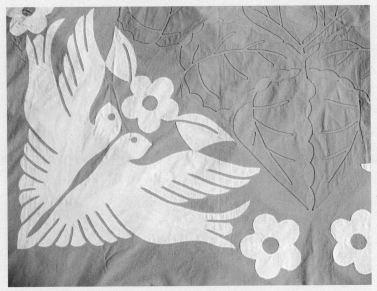

Tivaivai manu designed by Andrea Eimke and sewn at the Atiu Fibre Arts Studio.

Unwelcome birds

One day I visited a Rarotongan client who wanted to commission a new *tivaivai* as decoration for her bedroom. After discussing colours and sizes, she decided on a pattern that showed white terns.

By chance, I met her again many years later, while visiting Rarotonga. 'Do you know what happened to the *tivaivai* I commissioned you to make for me, the one with the birds?' she asked with a smirk in her face. 'My mother-in-law forced me to give it away! She said that birds on a *tivaivai* would be bad luck for a couple's bed, as they would cause the husband to *fly away with another bird*.' We both had a good laugh.

Some months later. I saw the *tivaivai* on a bed in an Atiuan friend's home, and I wondered whether my client had given it to her, and whether she had remembered her mother-in-law's warning when she made the gift...

triangular sections of patchwork *tivaivai* that were considered unsatisfactory in their construction, but which have been preserved as a cue to recall a pattern. Just occasionally, the repertoire of *tivaivai* motifs is widened to include an anchor, a kerosene lamp, a candlestick or birds. When enquiring into what is or is not permissible on the surface of a *tivaivai*, one is told of ill-fated designs that brought their owners bad luck and are avoided thereafter. Flowers appear, if anything, to be 'safe' and uncontroversial.

In representing the pliant, richly textured and brilliantly coloured plant kingdom, the Cook Islands *tivaivai* are reminiscent of the eighteenth-century collections of artificial flowers in glass, wax and silk preserved at the Botanical Gardens in Kew, or of nineteenth-century botanical drawings, themselves composites drawn from several exemplars, so as to capture the characteristic aspects of the plant in ways that would allow botanists to represent not just a plant species but an entire genus in a single image (Daston 2004: 226). Nowhere is the flower more evidently an expression of a way

of knowing than in Cook Islands patchwork *tivaivai*, where the flower, itself composed of a multitude of composite pieces, presents the viewer with a geometric abstraction of a perspective that commands knowledge of the internal workings of the plant, one that is invisible to the naked eye and that surfaces merely upon dissection. Like in a holographic projection, the pattern is then replicated in a fractal manner across the surface of the *tivaivai*. While the patched flower stands, as it were, 'hidden in the light', forcing one, in order to verify its identity as a type, to recall botanical

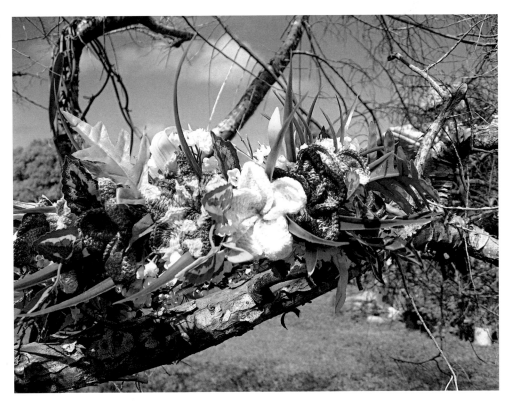

Crown of crocheted flowers.

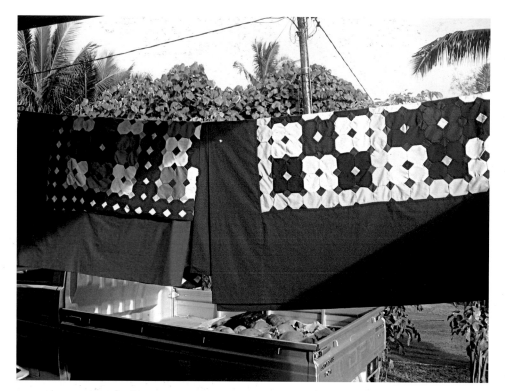

of shadow across the surface of the *tivaivai,* furthermore, creates the impression that the light source is not part of the field of vision, but originates from another dimension. It is these characteristics that are shared by Cook Islands, Hawaiian and Tahitian *tivaivai.*

There is thus something quite hyper-real about the flowers presented on the *tivaivai*, which appear to be there for a purpose other than that of mere representation. Like Marcel Duchamp's artworks, which experimented with ways of capturing the shadows cast by the fourth dimension (time) onto our three-dimensional world, Eastern Polynesian *tivaivai* appear to entice the viewer into contemplating the possibility of another world, one which intersects the visible plane via the creative channels of the mind. We will see in this chapter how – through the creative use of floral imagery alluding to the interplay of light and life, shadow and death, coloration and olfaction – concepts are brought to the fore that are as vital to the conduct of self and society today as they were two hundred years ago in pre-Christian times. These concepts concern the state of the souls of the dead, and their impact

knowledge acquired in the tendering of plants; the flowers in appliqué *tivaivai* elevate such hidden views to the surface by embroidering the shadings of sun-kissed blossoms and pollen with multicoloured thread.

The choice of colour for foreground and background, petals, leaves and pollen add to the impression that concepts are forged in the construction of the floral image which somehow capture what remains unsaid, and yet which inform actions vital to self and society. For the choice of colours and the manner of their superimposition is such as to create the impression of a light source situated behind the *tivaivai*, allowing the pattern from the forms of the cut-out fabric to stand in the shadows of another complementary pattern seeming to emerge from a bright light. It is this inversion of positive and negative spaces which, in a good design, should be well balanced and trick the untrained eye into reading the motif so that it appears to exchange foreground and background. The pattern itself is placed on a two-dimensional surface, yet appears strikingly three-dimensional, with the flowers' 'depth' giving the impression of full bloom. The even and regular distribution

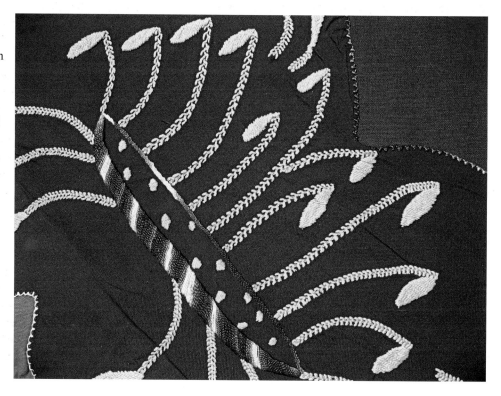

Embroidered stamens (detail of a *tivaivai tataura* owned by Ngakura Kautai).

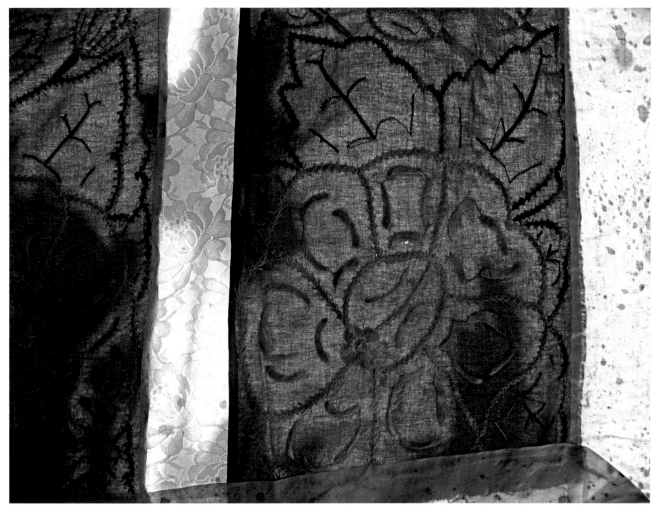

Light shining through a *tivaivai tataura* by Nga Mokoroa.

on the minds and conduct of the living; concepts which were then and are now intimately bound up with the creative transformation of cloth-like surfaces into shrouds for the dead and clothes for the living, and which met then and meet today with comparable cosmological ideas within Christianity.

Flowers have been said to speak a common language; so it is perhaps unsurprising that gazing at *tivaivai* one is easily reminded of other floral images from far away, such as those engraved on the surface of seventeenth-century Huguenot furniture. Flowers in the Huguenot diaspora served as a secret code, a hidden cue to memory, that in the diaspora had become vital for expressing visions of self and society in a way that would secure their transmission (Kamil 2005). Flowers, of course, are abundantly used in artworks across all ages and nations. The beauty of flowers is that they lack meaning and purpose, and thus appear to be depicted for incantation, rather than interpretation. Their only project is to flourish: growing, living and dying, soaking up the light whose presence they testify; attracting insects with their shape and bright colours, and 'feeding' them their pollen, offering them their pistils in order to aid their procreation. They are there, simply; almost as a kind of superfluous phenomenon in the midst of a complex world. No doubt it is the fragility of the flower, its testament to the presence of life and light, as well

as to its tending and transformation, which is presented to us within the folds of the *tivaivai.*

Larger than life, the botanical order created by *tivaivai* making in the Cook Islands incites us to rediscover the monumental and the complex in the miniature, the minimal and the simple, as leaves, petals and pollen project a world in close-up perspective. In the Cook Islands *tivaivai*, the size of the patchworks suggests a scale that forces one to step away in order to see the whole, yet this is in tension with the scale of the patched, and occasionally embroidered, floral motif, which draws the eye closer to the surface to identify the composition. The movement that is anticipated by these two scales could be argued to resonate rather well with the tracking back and forth from home that dictates both men's and women's lives in the Cook Islands. The micro-scale of the flower – represented either from 'within', that is from the perspective of the internal structure of a plant invisible from the outside; or from a perspective that draws the eye even further into a microscopic vision of the crystalline structure of its leaves and blossoms – effectively presents an iconic journey into the heart of Cook Islands life, coupled with an ironic realization that, hidden in the light, such journeys inevitably involve the presence of the foreign, the strange and the forgotten.

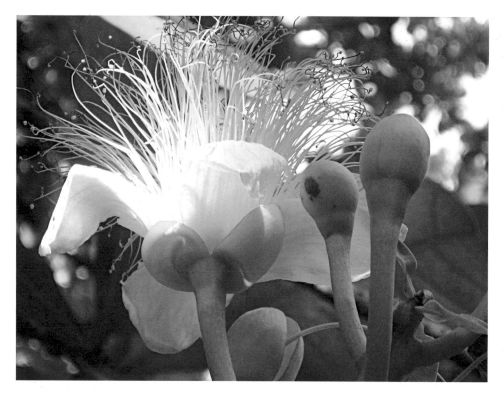

the tantalizingly trivial, extravagant and temporal nature of the flower that renders it a potent and lasting inspiration for the concepts that billow forth from the surface of *tivaivai*.

The tending of light

Tivaivai making, as it has been said before, involves the making of many choices. Choosing which flower to represent is just one of the many that have to be made, and it is one that is taken very seriously indeed. It is, in fact, a task Cook Islands women are well used to, as it is part of the daily routine of dressing. No woman would be seen in public without adorning her hair with at least one flower, if not with a flower wreath or a plaited hat made of coconut reeds[1] or pandanus leaves, worn precariously on top of the head. The Eastern Polynesians are unique in the Pacific for this emphasis on hair adornment, in which the scent of the flower, more than its size or colour, effects an olfactory halo that wards off, or attracts, complementary forces.

The macro-perspective at which the *tivaivai*'s over-all composition is visible is accessible only when they are draped over a support for presentation at competitions or ceremonies. The eye is then drawn not so much to each single flower, as to the symmetry that comes to the fore through the reduplication of the flower across the surface. There is a clear aesthetic preference for presenting this symmetry in ways that take experience and geometrically inclined thinking to come to grips with, and women take great amusement from testing each other's visual acuity in this way.

In general, however, the *tivaivai* is held close to the bodies of the women who engage in its embroidery and hand-stitching, much as it is held close to the extended family within which it circulates as a token of its internal relations.

The choice of flower, as well as the position it is given in the hair of the wearer, follows generally recognized rules. A flower in the hair or a flower *'ei* (garland) worn around the head or the neck is as much part of the island's dress code as the *pareu* shirts and dresses made from floral cotton. The colour of the flower is greatly associated with its scent, and

Touching the flowers that are arranged in precise order on the surface of the *tivaivai*, women often recall distant relations that are the key to their biographies, and who have required as careful tending as the flowers growing in the garden of the houses. Like the flower-cutting, the floral motif assembled on a *tivaivai* is humble, fragile and transient, and yet it figures as a testimony to the future of the family, which continues to have its roots in the Cook Islands. It is arguably precisely

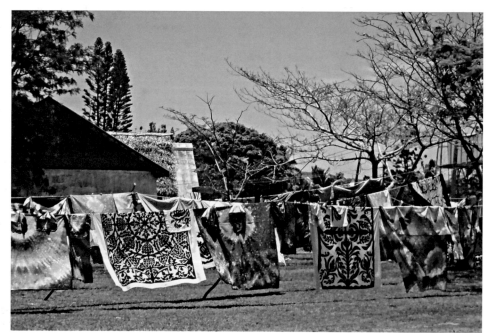

Tivaivai and dyed sheets displayed at the annual *tivaivai* show on Atiu.

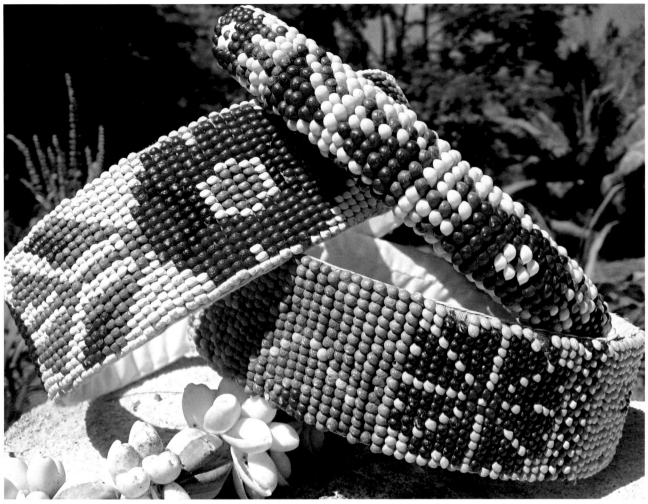

Hatbands made of tiny snail shells found at Mangaia and Atiu beaches and coloured with paint.

together they provoke associations that impute a certain intent to the wearer's actions. The scented flower on the head should be complemented by perfume, also made of flowers, and which is one of the Cook Islands most famous exports.

Long, abundantly decorated flower garlands or 'ei, some reaching down to the navel, are used to greet and welcome family and visitors arriving at Rarotonga International Airport. Likewise, boarding a plane to travel between the islands involves the donning of strongly perfumed flower garlands, which are worn as 'flying kit' by visitors and islanders alike. Such garlands are made of the delicately scented white flowers of the *tiare maori* (*Gardenia taitensis*), which was brought to the Cook Islands by early Polynesians when they journeyed here from Avaiki, and which is used in herbal remedies and in perfumed oils for hair and skin. Together with the *tiare,* a now less commonly used flower called *pua* (*Fragraea berteroana*; frequent on Rarotonga but quite rare on the other islands) is associated with the souls of the dead. *Pua* may have diminished in importance with the onset of Christianity, as on some islands the souls of the dead were believed to climb out on *pua* branches before leaping to the after-world (this tale is more Rarotongan than general to the Cook Islands). Drenched in the scent of the *tiare,* one cannot help but feel protected from the souls of the dead, a

protection which increases over the first few days on a strange island, when the decaying garland, hung near the door, proves too much for even the most experienced of souls.

Flowers in red or pink, such as hibiscus, gardenia, jasmine, frangipani, orchid and bougainvillea, as well as croton and cordyline leaves, are used for the adornment of buildings, floats and people, especially during festivals, as they are believed to attract those that may at first be rather dispassionate onlookers. The Tiare (flower) Festival, held over the first week in November and today a major event in the tourism calendar, is celebrated by a floral-float parade and a Miss Tiare Pageant, in which each contestant represents a different flower. Alternatives to flowers are head or neck garlands of *pupu* (land shells), which live on the upraised coral *makatea* of some of the southern islands, or on the various small lagoon atolls in the north. The *pupu* shells are dyed red, yellow and black, and are strung onto a single thread in a manner akin to the making of patchwork *tivaivai*. Precise counting of coloured shells is required in order to arrive at a consistent pattern when the string is tied in several adjoining layers around a circular ring, which is used to weigh down a Cook Islands *rito* hat. Practically speaking, the making of coloured hatbands with *pupu* shells has, however, died out. It used to be done with enamel paint that could be bought in small tins; but today, paints are classified as flammable goods for which one needs an import licence. Only one-litre tins are available now locally, far too much to use for hat bands, and thus they are considered unaffordable.

We cannot understand the central place taken by the flower in Cook Islands life if we think of it in purely representational terms. In fact, the similarity between the compositional structure of the stitched flower and the botanical drawings from the eighteenth and nineteenth centuries directs our attention to the possibility that what matters here is not just the symbolic capacity of flowers, but also their indexical function. The flower can be said to be an archetypal sign of an agency, such as light, that is captured and transformed into something else by the flower's own internal workings. We direct our attention to the flower not because it refers to a meaning that is independent of its existence (as in a symbol), but because it references an operational capacity to harness and transform light into colour and growth in ways that relate this particular flower to other flowers along a continuum of variation.

There would be nothing remarkable about the indexical qualities of flowers, which have long been cherished by botanists, were it not for the fact that the perception of the relations of colour, texture, petal, leaf and stem which make up the index requires us to plot them along a continuum according to a certain scale. All indexical relations exist in one scale or another, although it may not be very obvious to us why one scale is more popular than another at any particular time. Technology is certainly one constraining factor of scale; with the invention of micro-electronics and nanotechnology, for example, it is now possible to make things at and beyond the threshold of visibility with comparable functions to machines and computers which once filled entire rooms. Other constraining factors of the scale to which things are produced are the spatial relations between one thing and another. The scale ranges of common features of the home such as mattresses, for example, are in an obvious manner defined by their relation to the size of bodies, but also, perhaps more importantly, by the relation between the bed and other items of furniture required to be housed inside a room. The size of the *tivaivai*, for example, is that of a queen-sized mattress, which usually takes up an entire room inside a Cook Islands house. *Tivaivai*, in turn, may command trunks the size of a person for storage, or more moderate glory boxes, or even suitcases, in the outer islands, which in turn command the scale given to other furniture, notably benches that are kept in their vicinity.

We have seen in the last chapter how the scales in the representation of the stitched flowers are held to be

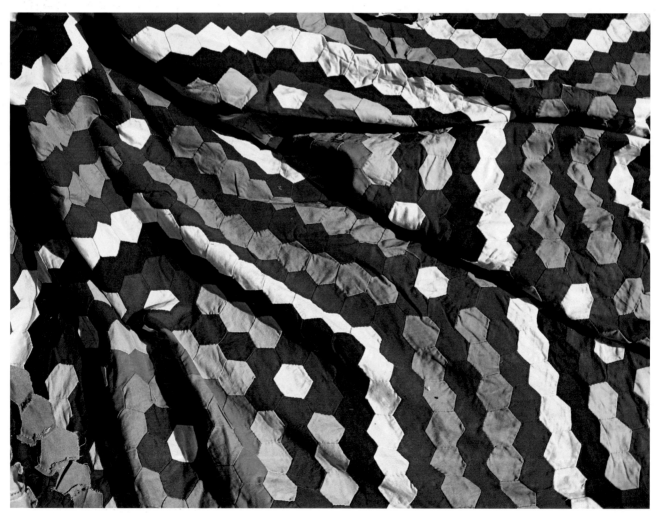

Tivaivai paka onu owned by Patikura Jim.

Making tapa

Tapa used to be made from the paper mulberry (*Broussonetia papyrifera*) or *aute* (*anga* on Atiu); two fig species: the banyan (*Ficus prolixa*) or *ava* (*aoa* in Maori) and the dye fig (*Ficus tinctoria*) or *mati*; and the breadfruit (*Artocarpus incisus*) or *kuru paea* (preferably, though any variety can be used).

The regular use and making of tapa probably vanished from daily life on Atiu early in the first third of the twentieth century. In his book *The Material Culture of the Cook Islands*, published in 1927, Te Rangi Hiroa (P.H. Buck) was able to present a detailed record of tapa-making techniques on Aitutaki, however it was already apparent that much of the ancient knowledge had been lost. There would have been few occasions, such as preparations for the investiture of Ngamaru Ariki in 1929, to see women still making tapa, because it was only intermittently produced, in rudimentary form, for costumes connected with traditional ceremonies and public performances (e.g. constitution celebrations after 1965).

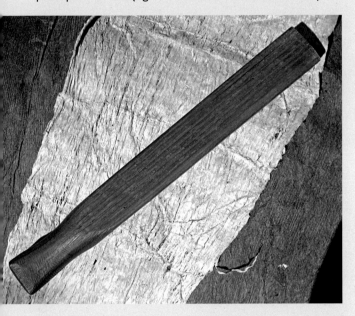

Tapa beater and paper mulberry (*aute*) and banyan (*ava*) barks; beaten twenty years ago.

Soon after my arrival on Atiu, I was able to witness preparations for the constitution celebrations, which were then held each year on the main island of Rarotonga. Each island would send a dance team to the capital to compete in a series of numbers that required traditional-style costumes. For the performances of legends and historical re-enactments, tapa costumes were generally used. I was told that Atiu and Mangaia were the only islands on which tapa making survived, and that after the celebrations were over other islands were keen on buying their costumes for ceremonial purposes.

Tapa was something I had never heard of before, and the new material stirred my interest. In 1986, on the occasion of the opening of the Atiu Fibre Arts Studio, the *mamas* (shareholders) and I invited women from the five villages on Atiu to join us in a tapa-making workshop. This was my first encounter with the techniques of tapa making.

My notes read as follows:

Use branches (of breadfruit, dye fig or mulberry) or aerial roots (of the banyan tree) as thick as a wrist, cut them to suit the size of your container (a 44 gallon drum in this case), fill half of your container with water, insert branches and roots, boil for ½ hour. Take out one branch and turn it upside down. When the fibres fold back, it is cooked properly. Take all branches and roots out, peel the bark off the wooden core. Separate the outer from the inner bark, wash the inner bark in fresh water and beat. Work all the bark towards one side, beating from the inside out, starting with the coarsest side of your *ike* (tapa beater). Working the bark back towards the other end, use the next finer sides of your *ike*. The beating is finished when a look against the light reveals an even distribution of the fibres.

If you don't want to use the entire bark right then, hang it up in the sun to dry and roll it up into a bundle like the pandanus leaves. Ventilate the bark every six months and keep it in a bag. Soak the dried bark for approximately one hour before beating to soften it.

The *mamas* were very patient with me, never tiring of my endless queries. Soon they voiced questions of their own, which we sought to solve by conducting extensive practical research into tapa making between 1986 and 1989, in which we tried out the various ways in which tapa could be made, guided by the knowledge that the local shareholders at the Atiu Fibre Arts Studio remembered.

commensurate with the scaling of kinship relations across time and space, enabling *tivaivai* to function as a mnemonic toolkit for conceptualizing biographical relations. This informational content of the *tivaivai* is enhanced by the fact that innovation is made only by *ta'unga* who have inherited patterns or special knowledge of design. Women who innovate patterns without such conferred rights to the knowledge of *tivaivai* are looked down upon ('*she thinks she is somebody*'), making it harder for younger women without a teacher in the family to take up patchwork.

Existing literature on concept formation has relied on the assumption that all concepts, and certainly those that extend to the ordering of social relations and of normative behaviour, are inherently complex and that they are like definitions, prototypes and stereotypes: essentially abstractions from belief systems that must be passed on piecemeal to the population as a whole. As socially effective ideas, we assume them to be free for all and to be shared as a common cultural good. A criticism of this view was made recently by the philosopher Jerry Fodor (1998), who argues that conventional assumptions about concepts have been seriously mistaken because they fail to recognize that concepts are about informational (mind–world) relations, that typical concepts are atomic and that they are carried most effectively by material entities which attract multiple associations. Cook Islands *tivaivai* make us think about how socially effective concepts are enshrined in material memory and thus are dependent upon the transmission of material knowledge in society.

Solving the last question

There are certain rules which need to be observed in order to honour the *tapu* of the *tutunga* (wooden anvil). Before (and after) using it, it is carefully cleaned. The *tapakau* (mats of plaited coconut fronds) are spread out, and the *tutunga* is placed on top, with bundles of dried banana leaves under either end, elevating it about 5 cm from the ground. Not only did this elevation enhance the sound, the bouncy leaf bundles also ensured that we would not damage our wrists in the beating process. Alongside the *tutunga* the *mamas* laid large *nono* (*Morinda citrifolia*) leaves to prevent our knees or feet from touching the anvil. Mami Tepu pointed out to me that we were not allowed to step over the *tutunga*, nor was eating permitted next to it. In the course of our research, however, I observed that those rules had lost their significance, and were no longer obeyed that strictly.

We tried out all the different ways of making tapa over a number of weeks, the *mamas* generously showing me everything they remembered and knew. We had not come across the forgotten secret of their foremothers, and still only ended up with narrow strips of bark that would not stay together when overlapped and beaten. 'You *papa'a* [white expatriates],' I was asked one day, 'you are used to learning from books. Can't you find a book that tells us how to make a large sheet of tapa without having to glue or sew the strips together?'

I took up their challenge and began to look for books on the subject, not an easy task in the days before the internet on a remote island like Atiu. With the help of friends, eventually I found a few, most of them giving me no better instructions than those the *mamas* could remember. But one book (Kooijman 1972: 50) mentioned something that we had not done: it recommended soaking the strips in water for 24 hours before beating, then washing them in fresh water to get rid of sap and any green matter that may still stick to the bark, placing them in a coconut basket to drain and then starting the fermentation process by 'wrapping the pounded strips of inner bark in banana leaves or coarse taro leaves (*Colocasia esculenta*) and holding them in this state for three days'. The fermented strips could then be beaten again individually until the proper thickness was reached. A continuous sheet could be formed by felting the overlapping edges of the individual strips, giving the sheet a final beating with the finest grooves of the *ike* and placing the sheet in the sun to dry.

I consulted Dennis, our Hawaiian friend, who confirmed that the Hawaiians also soaked the bark in the fresh water of their streams. However, Atiu does not have streams to soak the bark in. Nor does it have paper mulberry bark. So we filled a big pot from the water-tank, and submerged our bundles of banyan bark. Though the result was interesting, it was not at all what we had wanted. The bark was discoloured a burnt orange colour in some places, and light beige in others. Hmmmm...!? Could the iron of the tank or the aluminium of the pot have influenced the outcome? And then I had an idea.

Women on Atiu make dance costumes out of the bark of the 'tree hibiscus' (*Hibiscus tiliaceus*). The outer bark is first shaved off the sticks, taking great care not to damage the bast. The sticks are then bundled and taken to Avatapu, a small beach next to Atiu's harbour, where the bundles are buried under some of the heavy boulders of limestone that lie scattered on top of the reef only at that particular beach. The bundles are left in the sea for a week, the women hoping that the sea (or another woman!) will not steal their bundles. When the bundles are lifted out of the water, the bark has been bleached to a beautiful white and is easy to peel off the stick to take home, wash out the salt and dry in the sun.

In fact, Buck (1927: 79) even mentioned that the women of Aitutaki used sea water! So why not do the same? We prepared some aerial banyan roots by shaving off their outer bark, bundled them up and took them to Avatapu. We used the shaved-off outer bark as padding, and carefully placed the bundles under heavy

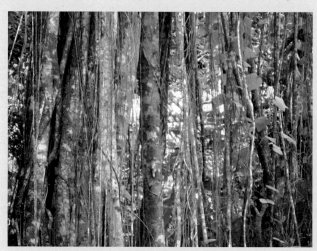

Aerial roots of the banyan tree (*ava - Ficus Bengalensis*).

Our attention is increasingly drawn, once more, to the associations and ideas inspired by materials and the way society is informing the production of materials and the transmission of the knowledge of their production. Once the driving force of the industrial revolution and the glue that bound together such prosperous families as the Dars and the Collinbrooks in the England of the eighteenth century, material knowledge was lost sight of as it was driven into laboratories. Hidden from public view and increasingly the domain of the expert, material knowledge continued for a while to dominate the world of women inside the Victorian parlour,

where they displayed the skills of sewing and embroidery, and their knowledge of diverse fabrics and materials (Parker 1984; Phillips 1999). At the dawn of the twentieth century, knowledge of materials and their iconic potential was to be finally displaced from the biography of persons to the expert knowledge of the designer and the industries surrounding the furnishing of the home.

There is no better testimony to the loss of the place of material memory in modern society than the paintings and collections of fabrics and costumes of Henri Matisse. Born in French Flanders in the textile town of Bohain-de-Vernandois

Two pieces of banyan bark have been joined by felting.

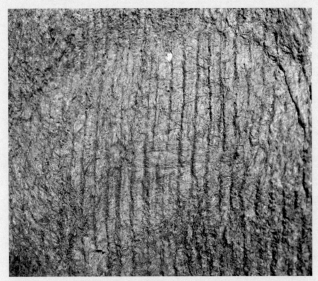

Banyan tapa with a different colour patch.

rocks that required several of us to move them. Then we needed one week's patience and a bit of luck.

When we returned, our bundles were still in the same place. But the bark had changed. It was of a beautiful light orange-brown and it came off the stick easily. Back at the Studio we rinsed it in fresh water and wrapped it in banana leaves. We gathered all our banana bundles in a *pareu* (patterned cloth) and kept them in a large aluminium pot with the lid on for three days.

When I was about to open the first bundle, the *mamas* teased me about the rotten smell that was going to come out of it. However they were wrong. The smell was indeed rather pleasant! The bark had a wonderful consistency, the fibres having loosened in the fermentation process, making it much easier to beat. In fact one had to be careful not to beat too strongly. I immediately made a trial, purposely choosing two differently coloured pieces of bark. I overlapped the beaten pieces by about 4 cm and felted the two layers together, carefully beating the overlapping area until all the bark had an even thickness again. Then we laid the two-coloured piece in the sun to dry. The result was great! Had it not been for the two different colours, one would not have realized the piece was composed of two strips. Today, almost 20 years later, that first piece still survives.

This technique now offered new opportunities. Wherever a strip had a hole, we were now able to patch it by felting a small square over it. When two or several long strips were joined, all our hands were required to handle the soft material. Strips could be added one after another, making as large a sheet as we wanted or were able to handle (limited as we were by the width of our *tutunga* and the many hands needed to move it from in front of the *tutunga* to the back). That day we all were very pleased that we had found the answer to our question.

Several years later I had a chance of visiting the Bishop Museum in Hawaii. In their archives they hold photographs from the beginning of last century that show Atiu *mamas* beating bark and carefully felting together the strips to make a large sheet. There was the proof that this was the method the foremothers had used. Mama Rangi even seemed to recognize her mother on one of the photos, who in those days had been considered a *ta'unga*, just as she was when we conducted our research.

Unfortunately, there was no longer any use for large sheets of tapa cloth, and the manufacturing process was too strenuous to keep up. The *mamas* returned to easier and faster methods, especially when they just wanted to make small items for the tourism trade.

Soon, Atiu gained fame for its tapa flowers, both single ones and the entire *ei katu* (flower-wreaths worn on the head) that can still occasionally be found in the gift shops on Rarotonga.

in the second half of the nineteenth century, Matisse grew up in a family of weavers, and it was the opulence and delicacy of colour woven into fabric which proved a lasting inspiration for his paintings, his drawings and the semi-abstract stained glass and cut-paper compositions of his later years. He owned a large textile collection and decorated his home and studio with bits and pieces of fabric he found in junk shops, battered unmatched fabric-covered chairs, faded textile hangings, threadbare carpets – 'noble rags', which formed the 'real life partners of a luxuriant imagination.' (Spurling 2005: 15).

Matisse's flight to the south from the drab and confined world of the northern industrial city of his birth, and his search for the richness of optical effects incurred by light, was governed by his familiarity and passion for textile, and it is no coincidence that this pursuit of ways of recapturing the enchantment of childhood led him in the last few years of his life to French Polynesia. For in the 1940s sewing *tifaifai* was as culturally prominent in Tahiti as making *tivaivai* is today in the Cook Islands, and it inspired his gouache collages and his stained-glass windows. In fact it has been proven by John Klein, in his research into Matisse's correspondence with Pauline Schyle, that Matisse owned two *tifaifai*, although they are not listed as part of his collection (Klein 1997: 58–9). Matisse documents the influence of *tifaifai* on his paper cut-outs in a note on the back of a photograph he sent to Mme Schyle in the 1940s (see Klein 1997: note 54). Klein also refers to the unpublished research paper by Angela Levy (University of Pennsylvania 1987) which establishes interesting connections between Matisse's cut-outs and Tahitian *tifaifai* (Klein: ibid.). It seems clear that Tahitian *tifaifai* influenced Matisse's interest in decorative art whose composition was not merely determined by a natural setting, but which would rather create its own environmental effects. We admire Matisse's paintings as they present us with a lost sensitivity towards the material and a world where socially relevant associations and ways of doing things are carried by materials. It is fitting that we should follow Matisse in enquiring into the concepts that are provoked by material icons in Eastern Polynesia – concepts that came to be expressed in working with pieces of fabric and with the iconic potential of flowers. Material memory was as important for Matisse as it is for the people of the Cook Islands for reasons that we may want to take the trouble to understand in order to see how it creates a difference to the way people live their lives.

Luxurious works with fabric such as patchwork *tivaivai*, perhaps as much as a well-tended garden, open up a world in which the significance of things goes beyond the 'merely symbolic' or 'merely functional'. Nowhere else is the value attributed to fibrous things more openly declared than in Polynesia, where cloth-like artefacts such as mats or barkcloth have been documented by numerous ethnographies not just to comprise a form of wealth; their possession also sanctions political succession to office and land ownership tied to genealogical position. While fibre-arts may thus be said to be iconic in expressing ideas and values across Polynesia as a whole, imported cotton fabric does not, however, resonate in the same manner everywhere. In fact, interpreting coloured cloth as a material icon in ways that provoke its cutting and re-stitching into the shape of flowers, as happens on the Cook Islands, has not become a pan-Polynesian tradition, despite the prevalence of photography and women's workshops that carry craft-knowledge across the entire Pacific.

We can liken the spread of a disposition towards 'material iconicity' to the distribution of tattooing in the Pacific, which similarly is not amplified everywhere in the same manner, and differs in the extent of application and in its interpretation from island to island. The anthropologist Alfred Gell (1994) has shown in his *Wrapping in Images* how the epidemiology of tattooing closely follows different political articulations: the greatest amplification of tattooing is present in the largest islands which permit the most pronounced articulation of spatial and social distance between commoners and aristocrats. Given the fact that both the surface of the skin in tattooing and the surface of cloth in patchwork are worked into patterns that serve to contain power within the body, the analogical treatment of the body's surface and the surface of cloth would prompt the hypothesis that the spread of patchwork in Polynesia may follow closely that of tattooing.

Unfortunately, ethnography is far less explicit about the extent and presence of patchwork than it is about the practice of tattooing, but it is clear that in Fiji and Samoa, where tattooing is interpreted as a process of wounding involving an exchange of blood that creates relations between subsequent generations of siblings, patchwork is absent. On the other hand, where the tattoos are interpreted as an intimate part of cosmology, such as in Tahiti, where the sacrificial offering of the tattooed skin to the god *Oro*[2] formed part of most rituals in pre-Christian days, patchwork is present. From this simple observation, one can conclude that a belief in the agency of the decorative surface of body and cloth is necessary for patchwork to take off as a social practice. As Nicholas Thomas (1995: 110) pointed out, 'just as men created a person in the world through the addition of artificial skins [in the form of tattooing], women created a person for the next world by undoing those skins by stripping away the armour through which the warrior [a person acting as mediator between the living and the gods] had been constructed.' The removal of layered *tivaivai* from the grave at the unveiling ceremony (see Chapter 2, pages 25–6) effects just this.

One look at patchwork *tivaivai* is sufficient for us to notice that its patterns emerge from within a sea of colours that are vibrant and polychromatic, their intensity heightened through careful juxtaposition of cold and warm colours. Such an emphasis on coloration may seem particularly odd if one is familiar with the treatment of tapa cloth from Western Polynesia, which to this day uses black, brown and white. Yet patterns in colour are nothing new; collections of mats and barkcloth made in Eastern Polynesia, and especially in the Cook Islands in the late eighteenth and early nineteenth century, show an emphasis on coloration that included deep reds and browns and black tones as well as yellows. Famously detailed descriptions by the Maori anthropologist, physician and politician Sir Peter Buck, also known as Te Rangi Hiroa, capture the material culture of the Cook Islands and of Tahiti in the mid-nineteenth century and describe to us the care and attention devoted to creating multicoloured cloth surfaces. The son of William Henry Buck and Ngarongo-ki-tua, a Ngati Mutunga Maori tribeswoman, Buck was a medical officer for Maori health with the New Zealand health department, and as such was stationed in Eastern Polynesia. He describes the arduous production of coloured-fibre cloth as lasting many weeks and months. The inner barks of certain trees were stripped and soaked in water before being worked into lattice or beaten into sheets of fibrous material which were then stained or painted in a process even more time consuming. Cook Islands barkcloth was made in large and rectangular sheets, some of which are bright yellow on both sides and measure 2.92 × 1.72 m – the size of a latter-day *tivaivai*. (Buck 1944: 72). Other pieces are dark matt-brown on one surface and have a shiny almost black appearance on the other. This form of decoration was achieved by soaking the cloth in the mud of the taro swamp, washing and drying it, and rubbing the other side with a mixture of grated turmeric and grated coconut, in part to perfume the cloth.

Akapareuanga: distributing textile wealth

The word *pareu* formerly described a piece of tapa that men and women would wrap around themselves (Savage 1980: 236). In modern times it has come to mean a printed cloth (Buse and Taringa 1996: 322), so typical of most South Pacific countries, that displays tropical flowers or western Pacific tapa patterns.

Aka refers to the dance the women – who proffer these textile gifts to newly weds, haircut boys, birthday children and dignitaries – perform to celebrate the joyous occasion of forming a new bond between two individuals, and thus between two kinship groups, on occasions such as coming of age or being honoured as a dignitary. The suffix -*anga* describes an action.

Typical printed 'island fabric' locally called *kakau pareu*.

Filling a dowry chest

After wedding gifts are presented to the young couple (by members of both extended families), the respective family's speaker will call out what it is, to whom it is destined to go and who has given it. All gifts are meticulously recorded in the family's record book and bring with them the obligation to reciprocate with a similar gift in a future *akapareuanga* held for a member of the giver's family. Whether it is a selection of shirts for the groom or dresses for the bride, bed sheets, 'minks' (soft synthetic blankets brought from overseas), brightly coloured satin quilts, many metre-lengths of printed fabric or precious *tivaivai*, all of it is unfolded and held up by at least four women. Accompanied by loud cheering and singing, these presenters will then dance up to the young couple, waving the presents up and down, and will either wrap them around the individual or couple, or lay them at their feet. Depending on the wealth the families wish to display in their lavish feasting, the newly weds can sometimes be covered up to their chins in 'textile wealth'.

The use of textiles is not accidental. Cloth and threads are used in nearly all cultures and civilizations as a metaphor for society and social relations, connecting past with present, one with the other.

The textiles given represent the ties that bind both the husband to the bride, the young couple to their respective families and the two kinship groups together, all as a result of the wedding ceremony. These ties will commit the 'recipients to loyalty and obligation in the future' (Weiner and Schneider 1989: 3).

A marker for important stages in life

The *akapareuanga* can be a part of hair-cuttings, twenty-first birthdays or other celebrations of major stages in a person's life. It can also be a tribute to a dignitary. Late in 2006, the current Parua Ariki, who has never resided on Atiu since his childhood days, came home for the opening of the palace which his tribe, the Ngati Nurau, had built for him in the village of Mapumai. A grand feast was prepared. The bright full-moon night preceding the big day was filled with the squeals of pig slaughter, the smell of smoke and cooked food from the *umu* (earth ovens) and the scent of flowers strung up into garlands. Members of a Tahitian branch of the Nurau tribe had arrived the day before the inauguration. On the palace's veranda, the chief's seat was decorated with *tivaivai* and pillows, and the lawn leading up to it was adorned with flower arrangements stuck to banana trunks. Ribbons were cut, speeches were held, the palace was officially opened and the chief honoured. When the Tahitian's turn came, the members danced forward, the group's speaker at their helm and, singing traditional chants, they presented their ariki with *tivaivai* and an enormous length of printed fabric, waved about to mimic the sea across which they had flown to attend the celebration. They then wrapped him in intricately sewn *pareu* and placed a selection of pots and pans at his feet as a contribution to his new household. The ties across the ocean and the boundaries of two different nations were once again strengthened by these symbolic gifts, which at the same time connected those present with their common traditional past.

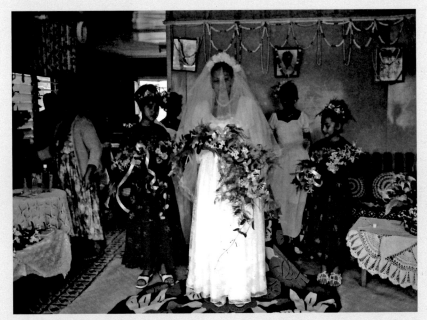

A bride on a *tivaivai*.

In the part of Polynesia where the staining of fibres appeared to be of great significance, sewing has today generally replaced barkcloth making, as precise quantities of coloured cotton fabric are cut and shredded into pieces before being stitched into intricate and vibrant floral patterns of greater or lesser abstract appearance. Cotton fabric in fact appears to have answered an interest in capturing the material effect of light, which was common throughout Eastern Polynesia. Emanating from darkness, pre-missionary cosmogony has light accompany persons in death in the form of colourful cloths (today *tivaivai*) that are wrapped around the body in the grave. Prior to the introduction of cloth, shiny items such as shells and feathers were attached to cloth-like surfaces, like feather headdresses, to heighten luminosity.

The reinforcement of reflective qualities turned luminous and coloured things into an effective mediator with the immaterial world. White, ready-patterned, coloured or sometimes even odorous fabric offered new technical possibilities to harness, contain and control the luminosity of significant decorations. By sheer coincidence, the fabric which Europeans brought to barter, together with glass beads and mirrors, was often of a vivid red or blue, as the Europeans' idea was to attract people by the colour and dazzle of their wares.

White cloth is generally assumed to reflect light most effectively, both to hold powers within and to ward off external forces, and is thus considered the most appropriate colour for clothing worn in situations of sanctity in which the powers inside the body are in danger of escaping. Coloured cloth, however, is associated with the making and sustaining of connections in the material world through its refraction of light. The capacity of colour to evoke connections with distant worlds is sensed today when looking at *tivaivai* in the diaspora, where the associations and memories which colour provokes may be enhanced by the muted light in which *tivaivai* come to be seen.

Across Polynesia we thus find that cotton fabric is used, singly or in combination with fibre products such as barkcloth or mats, to harness light and to make it do its work. There are, however, some interesting regional differences in the effects attributed to materials that exude colour, creating a lasting difference in attitudes to the take-up of material technologies such as sewing. In Western Polynesia, neither missionary activity nor the arrival of new types of cloth interrupted the manufacture and use of indigenous cloth. 'Cloth' was and is 'barkcloth' made from beaten strips of mulberry bark, or it consists of woven material made from dried strips of leaves or pandanus fibres, both materials grown close to the house. The bark is treated so as to appear light, with dark patterns being imposed by stencilling or rubbing; the leaves of the pandanus are softened and bleached and subsumed into the weave of the mat. Such cloth of bark or leaves ('tapa' and fine mats) still serves to wrap people of rank and completes a presentation of wealth by being placed over the top of a pile of other ceremonial gifts. It is almost as if white, store-bought cloth is thought of as still coloured, or as not fully drained of colour, or as luminous and reflective of light as the fibre that has been whitened by the sun. Quite in contrast to the whitened mats and barkcloth worn by those of highest rank or in situations that require the harnessing of powers within the body, everyday clothes in Samoa are of vibrantly coloured printed cotton or, for very formal occasions (Church or political meetings), of a dark fabric without any printed patterns (Tcherkézoff 2003: 53). Here, it is almost as if decorated cloth is not required to maintain sanctity, which in Samoa is in fact managed by the tattooing of the skin underneath clothing; instead fabric serves the purpose of differentiation, with the distinctions between bleached fibres and coloured cloths articulating with those observed between individuals and in society at large.

In Eastern Polynesia, on the other hand, the destruction of clothed wooden sculptures by members of the London Missionary Society halted the ritual use of barkcloth and fine mats forever. Here, in a move that was as swift as it was seemingly unremarkable, techniques of cutting and sewing liberated the work that went into the patterning of layered barkcloth funerary cloaks and into the fine mats and protruding feather holders (intricately knotted into the shape of flowers) with which genealogical figures (the so-called staff-gods) were covered. The rapid take-up of ready-coloured cloth in this part of the Pacific suggests that coloration was attributed an efficacy that only absorbent cotton could guarantee, with whiteness, worn here regularly to Church or to important meetings, assured equally through the draining of pigmentation from cloth surfaces. An efflorescence of pattern innovation in the ready-made medium of dress-material and store-bought plain-coloured cloth thus came to shape the development of *tivaivai* across the Cook Islands, the Hawaiian Islands and the Society Islands.

Where patchwork prevails, colour is thus regarded as a material substance, one that is indicative of the performative, fractal properties of light that has escaped its container. It thus appears entirely logical and persuasive that this material substance should be harnessed by cutting and by binding it back into a surface that in its floral imagery recalls light's potential to enhance growth and life. Fabric is a material resource which possesses both a spiritual and an aesthetic dimension. Stretchable and foldable, it can be wrapped around the body as well as stored in containers. More than any other material, however, fabric is a skin-like surface that responds to inner and outer change and induces multi-sensorial perception: it registers presence in its imprints, when marks are stencilled, blotted or painted onto cloth; or invokes absence through staining, ripping or crumpling; it serves as an archive of use and of value through its manifold qualities of texture, sound and smell; or resonates ideas of continuity through the work that goes into the sculpting of its architecture in sewing and tailoring.

Cook Islands patchwork, however, thrives not just on the potential of fabric, but also on its verisimilitude with flowers. Just as flowers are what they are because they harness light and turn it into growth without us ever being able to see this quasi-magical act of conversion, Cook Islands women like to display rather than demonstrate their control over the application and transmission of material memory. Although many women in these islands shoulder tasks such as rising before dawn to work at the airport before the first plane arrives at 4 a.m., returning to the village to work in the vegetable garden and cook for often many children and relatives, all the while continuing to plan and execute sewing, and possibly holding down other offices in the community, it is considered rude to talk about work. In considering how to take on life, Cook Islands women are guided by the multiple associations that are harnessed by the patched and embroidered flower which imprints itself on the mind as a lasting vision of self and society.

Olfaction and transition

Cook Islands *tivaivai* are perhaps the most complex of Polynesian *tivaivai*. There are, as we know, several distinct named types, each composed of up to three layers constructed from pieces of cloth and cotton thread. It is, as we have also seen, to the decorative dimension of the *tivaivai*, their visual and conceptual complexity, that we must look in order to understand the ideas they carry and the difference

that these have made to Cook Islands society, wherever its people live.

When we look more closely at the decorative surface of *tivaivai*, we notice three features that appear strikingly at odds with one another: firstly, as already mentioned in passing, the flowers that are represented are not commonly found in Eastern Polynesia. Secondly, the flowers that are depicted on *tivaivai* are magnified, highly naturalistic representations that tend to visualize what is usually invisible to all but those who love flowers to the extent of wanting to assure their exact replication; what we look at are in fact graphic representations of dissected flowers, split open to show the working parts of the plant. What we see on the surface of the *tivaivai* are the capillaries and stamens of flowers from an unusual 'inside' perspective, which more frequently than not makes the identification of the flower a difficult task. The third feature is perhaps the least obvious, but also the most pertinent to the iconic potency of flowers; for represented are not flowers standing in full bloom, but cut and dying flowers that are as carefully arranged as the cuttings that decorate the flower headbands without which no women will be seen

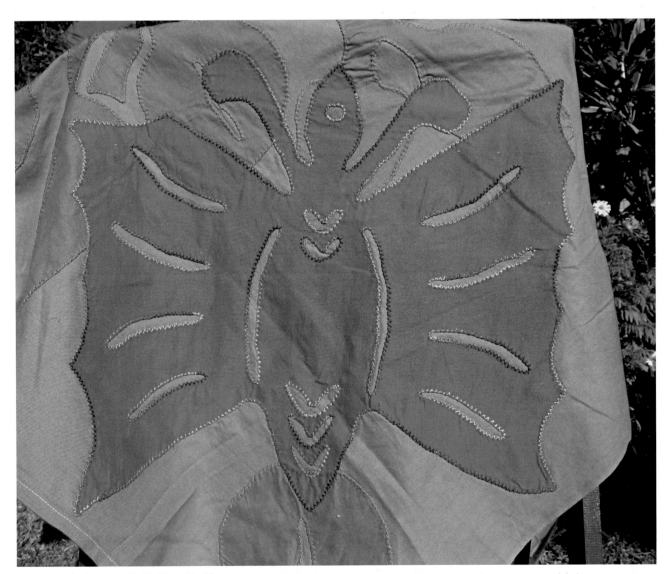

Tivaivai manu karakara owned by Tira Tararo (detail).

in public places. It is dying flowers, in fact, that have the most potent olfactory capacity.

Taken individually, each of these characteristics appears quite unremarkable. We have already looked at the role of foreign flowers, and have remarked on their importance in drawing attention to trade and to migration, which have dominated the Cook Islands historically and in the present day. Flowers have been described by Ruth Phillips (1999) as the principal iconography of 'trading identities', and thus as an icon of the most potent and yet also fragile connections that link those who have departed with those who stayed behind. Ruth Phillips refers in her work to the trade that linked Victorian England and the Northern Iroquois tribes in the late 1900s, yet her insights can well be seen to be relevant to Eastern Polynesia. Like the common humanitarian concern with death and its social impact, the flower was the material icon most in common for Chinese traders, European explorers, missionaries from the old world and the chiefly class in the Cook Islands, the latter of whom adorned their heads with elaborate and tall contraptions decorated with tightly knotted floral-shaped holders into which feathers could be inserted. The flower, no doubt, has been significant in indexing status and power in the Cook Islands for a long time, with foreign flowers speaking overtly about the means people have to encompass the outside world. Like the richly embroidered mementos that late-Victorian ladies brought back from their holidays in North America, made by Indians to suit their taste, the flowers which decorate the *tivaivai* of Cook Islanders are also trophies of exploits and discoveries. Certainly, floral prints decorating dress material imported by missionaries from England were popular in the early part of the twentieth century, when travel to other parts of the world was still relatively constrained. Another source of foreign flowers depicted on fabrics prior to the widespread onset of out-migration and travel was the presence of American GIs on the island of Aitutaki, where the American air force constructed a landing and refuelling base. The large lagoon of Aitutaki had already been used as a landing-site for sea-planes which had operated between New Zealand and Tahiti.

We also have already discussed the practice of presenting flowers cut open or dissected, revealing, as it were, the inner operating mechanism that turns the flower into a light-processing and transforming agent. The agency of flowers can be said to be further enhanced by the tendency to depict them in motion: rotating around their own axes in step, often turning back upon themselves, with their duplicates, all mounted on the planar surface of the *tivaivai*. There is nothing passive or fixed about the motifs we see on *tivaivai*, which appear to animate the cloth and make it move in front of our eyes. This movement, often clockwise in the appliqué *tivaivai*, is unique to Cook Islands patchwork and can be said to be one of the 'trademarks' that distinguish it from those made in Tahiti or Hawaii. The flower as the location and agent of growth, of movement and of change captures a notion that is deeply embedded in the Cook Islands consciousness, which sees these islands as the literal turning point of Polynesian history, a place of transition and a vital stepping stone in the conquest of the Pacific Ocean. It reminds one of the Cook Islands' geographical and political position: in between French Polynesia and New Zealand, two poles which connect Cook Islanders to their past and to their future, and which have done so since historical times when seven great canoes, arriving from Tahiti and the outer islands, set off again from the beach of Avana Harbour in Ngatangi'ia on the island of Rarotonga to populate Aoatearoa/New Zealand, an event which is believed to have taken place in AD 1350. One can go to Avana Harbour today and see the spot where the canoes departed, now marked by a circle of seven stones, commemorating the famous seven canoes: Takitumu, Tokomaru, Kurahaupo, Aotea, Tainui, Te Arawa and Mataatua, all names of present-day Maori tribes.

Located within a small triangle of the Pacific defined by Fiji, Samoa and French Polynesia, Cook Islanders' ties today are as close to New Zealand, Sydney and Los Angeles, where three quarters of the Cooks' population live, as they are to Tahiti. Like the foreign flowers that decorate Cook Islands *tivaivai*, prosperity is found elsewhere yet reaches the Cooks along the paths stitched into the patchworks. The movement which the flowers on the *tivaivai* capture as they rotate along their axes is indeed part of everybody's life in the Cooks. Every year, local churches organize trips to neighbouring islands, but also to far away places like Australia. It is for such visitations that *tivaivai* and other special Cook Islands floral paraphernalia, such as the flowers made of tapa-cloth on the island of Atiu or the 'cotton flowers' of Mangaia, are busily prepared during the months preceding such journeys. In addition to such collective visits, women also travel individually to relatives abroad or on neighbouring islands, accompanying the scouts on their trips to China or other such exotic locations, or on holiday to Hawaii. The contrast between the hours of sitting still sewing and the hectic life spent travelling to and fro is perhaps best revealed in the visual tension between circulation and stasis that defines the symmetrical arrangement of flowers on appliqué *tivaivai*. Travel is not just an occasional event in the lives of Cook Islands women, but a daily necessity, as the customary transmission of land means that people acquire curiously divided chunks of property. A family can end up with a house by the coast, a pineapple plantation somewhere else, a taro and vegetable garden on the other side of the island and the odd group of papaya and mango trees dotted in isolated patches here and there. Scooters are essential in the Cook Islands, even on the smallest of islands, and days are dissected into purposeful trips around the island.

From all this we can see how the depiction of flowers on the surface of *tivaivai* resonates rather well with ways of being and thinking in the Cook Islands. But what about the cut flowers, presented occasionally in European-looking bunches or baskets? We know that cut flowers are considered essential decoration for women's heads and hair and for those who are arriving or departing. Heavy garlands of flowers, worn reaching down to the navel and already wilting soon after construction, almost suffocate the wearer with perfume that is so strong that it appears inconceivable it could emanate from the small flowers (strung up in a fashion reminiscent of the

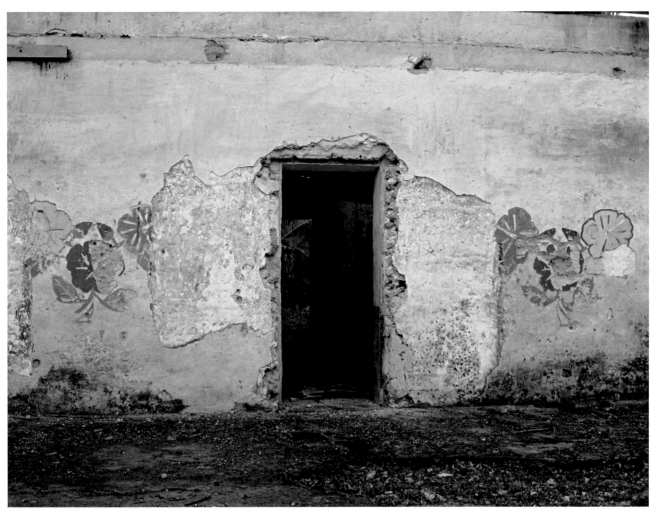

Empty house on Mangaia.

construction of a piecework *tivaivai*), and which gains further strength as the flowers decay. There can be no doubt that cut flowers are productive of smells that living flowers can only indicate as potential.

Olfaction is the sensory mode that induces memories of objects or persons more powerfully than any other. The smell of a lemon, for example, conjures up its image, and even makes one's mouth water as one identifies the smell, even if it is invisible. If you have never seen a lemon, cut it open and tasted it, you will not be able to identify the odour. Odour can thus induce a remembrance of something that is no longer there, and its power for provoking recollection is what appears to be captured by the cut and embroidered flower, which, although it clearly does not engage the same sensory register, does in its visual evocation of the transition between life and death, presence and absence exploit a material resonance with olfaction.

Floral imagery, when presented as severed from its roots, gives powerful expression to the experience of transition and loss, signs of which are visible everywhere on the islands. Especially on islands such as Atiu or Mangaia, which have seen the largest out-migration, boarded up or abandoned houses, and abandoned gravestones where houses once stood, are common sights. Talking to the women and the old who

are left in the remaining occupied houses, one hears tales of sons, fathers and uncles who have left the island, some as recently as ten years ago when labour migration swept many of the able-bodied men to the far shores of Australia to help in construction for the Olympic games there, most never to return. Life is hard for those who stay behind, sweetened only by occasional reunions at weddings and funerals, when tokens of lives on far away shores may be received.

The cut flower, so close to death and yet still so much part of everyday life, is an apt symbol for the transitory space which death occupies in Cook Islands villages. There are hardly any houses that do not have a grave on their front lawn, usually just outside the veranda or the entrance door, where women tend to sit and stitch their *tivaivai*. These grave-structures are large, and resemble the house of the living behind them, with foundations, platform and raised roof-structure (see photo page 25) Although occasionally fenced in, these are not sacrosanct spaces, but places that are fully integrated into domestic life – washing may be hung up to dry beneath the roof and flower-pots are likely to decorate the platform beneath. Entering the house from the front, one inevitably walks past the grave. We may see this as a memento mori, a reminder that life is transitory, brief and uncertain, yet this is probably not the case at all. A rather different sense of the central place occupied by death emerges from newspapers, where death days are memorialized by large columns of announcements, alongside those of birth and weddings. The departed, some of them having died sixty years ago, are recalled here with attention to detail: alongside a list of the names of families that have descended from the deceased, are pictures and events that evince a personal touch.

The dead, we realize, have not fully departed, but continue to share a transitory space with the living for a number of years. Their death is recalled annually in ceremonies that remind of the old pre-Christian ritual of *pa'iatua*, the ritual gathering of the gods, which the Cook Islands shared with neighbouring Tahiti. This ancient ritual had three stages, mirroring the three cycles of the agricultural calendar, each being defined in relation to the manipulation of a mummy-like object:

1 the unwrapping of the object, effecting the death or departure of the dead;

2 the exchange of feathers and of cloth as the 'sharing' of the remains of the dead;

3 the re-assemblage or 'renewal' of the object, invoking the return and emplacement of the dead and the period of abundance.

The *pa'iatua* ritual established a contractual relation with the multitude of the dead, and secured their annual return. Today, remnants of the old ritual are associated with burial, and the first anniversary of the death, when a second stage takes place as the grave is adorned with a headstone. Mourning is officially ended at this second stage, as the deceased person's official titles and land are passed on to the next generation. Thereafter, yearly recollections, marked by the family on the anniversary of the death, take place for a period that may last as long as another life-time. Only with the final anniversary, when the years after the burial equal the age of the person at death, is the person allowed to sink into the collective memory of ancestral connections, which connect the living with an apical ancestor.

The intermediary stage of personal or social transition is classically described in ethnography as being marked by the removal of structures of containment, which symbolize self and society. In some societies of the Pacific, initiates on the threshold to adulthood leave the village and undergo rituals of body alteration through tattooing, fasting or, conversely, the ingestion of special foods. The anthropologist David Howes (1987) has argued that societies that extend this process of transition to the treatment of the dead associate the transitory state of the soul with odour, the experience of which as a material substance is most heightened at the point it escapes its container. The emphasis on olfaction, which we find in the Cook Islands and which underlies much of its visual and material culture, is thus a commentary on a cultural conception of the soul, which considers its articulation in living minds to flourish best in situations that complement the transitory stage of the soul after death.

The connection we can begin to see here between olfaction, transition and the state of the soul of the dead, or the mind of the living, explains one of the most important customs of the Cook Islands, which supports the transmission of knowledge critical to society from generation to generation. Women, for example, who are known as originators of *tivaivai* patterns are considered *ta'unga*. This term is usually conferred onto ritual specialists and designates the special status afforded to these women, who stand out also as superior organizers of public functions, acting in roles such as a president of a non-governmental organization or a leader of a youth group. What these women tend to have in common, however, is not their very public persona, but something we might consider intensely personal. For rather surprisingly, the designers of *tivaivai* tend to have been adopted across tribal boundaries. To take on a 'feeding child', as an adopted child is known, is a common practice in the Cook Islands and is undertaken most commonly by a couple whose own children may already have left the house, by childless relatives or by the child's grandparents. The 'feeding child' is taken into the house, cementing connections between households and tribes.

'Feeding children' are usually the first-born male or female grandchild. The child is a 'cutting', and the knowledge that is conferred upon the feeding child about relations of land, labour and loyalty in the form of floral patterns is not individuated, but retains the plurality of the tribe as an aggregate of personhood. For the feeding child 'impersonates' an assemblage of relations, collapsing the singular with the plurality of the persons comprised by the lineage. Like the staff-god, the notion of personhood conferred upon such a child consists of amalgamated replications of itself and it is this internalized self-similarity that is carried forward into the *ta'unga*'s creative ability to replicate relations externally into the cutting and sewing of patterns.

The sewing 'proper' of *tivaivai*, tablecloths and pillow-covers is usually started only upon marriage, or sometimes even only when a woman's children are ready to leave the house. More often than not, the first sewing will be prompted by marriage into the spouse's family, when a wife is expected to provide a *tivaivai* to be placed into the grave. Every *tivaivai* made to be given away elicits the expectation of a similar return when a woman's own children get married or when there is a death in the woman's own close family.

The giving and the receiving of *tivaivai* connects those who may be living apart for most of their lives, but they also demarcate moments of rupture and the start of a transitory cycle, when relations are transformed forever. In more than one way, each new *tivaivai* anticipates a new transition and a new severing of existing relations, so that new ones may emerge. We may think of these sewings as skin-like, self-replicating and layered cloth-things, whose successive opening and folding traces the frequent departures and the many returns that mark Cook Islands life. Marking points of departure and new beginnings, the sewing is a shroud that conjoins persons in death as its ties together those that are divided by fate and circumstance in life.

As over the years new backings are sewn onto old stitching, a woman settles to sewing with thoughts in mind of those whose death will conjoin their soul with her state of mind. As she faces the house of the dead just outside her doorway, the passage through the household that she has anticipated and called forth through her work takes on another meaning. No longer does it appear to lead from an inside to a world outside, but from a place of plurality and public responsibility that is synonymous with the inside of the house to a place of 'oneness' where the *many* social relations that are recalled by the *tivaivai* are transformed into *one* as the *tivaivai* are wrapped around the corpse inside the grave. And no other material icon can provoke associations that bind the material world to the immaterial world of the soul as well as the coloured flower, whose fading light brings to the fore the magical power of olfaction.

The sense of things: on the sensorial relations between sewing, dancing and singing

The cultural elevation of olfaction to meaningfulness associated with the mediation of the states of the soul and the physical and the social body rarely occurs on its own. For the realization of olfactory associations with material things is a classic indicator of synaesthesia, which appears in language and in perception, and constitutes the relating of languages of different sense modalities and perceptual experiences, conflating seemingly disparate sensory realms, such as pitch–brightness, colours–shapes, sight–taste, taste–smell, smell–hearing or hearing–touch. In the case of the Cook Islands, smell is associated with the brightness of colours and the pitch of colour terms,

Tivaivai ute

In 2007, for the Cook Islands Tivaivai Associations' annual *tivaivai* exhibition at the National Museum on Rarotonga (usually held from mid October to mid November), we decided to invite all Cook Islands artists to exhibit works inspired by *tivaivai*, in whatever medium they desired.

The Takitumu dancing team performs *ute*. Photograph courtesy Ewan Smith, Rarotonga.

The *ute* that the Rarotongan village of Takitumu's dance team performed during the 2007 constitution celebration had won first prize. As it was about the introduction of *tivaivai* to their village, we invited the Takitumu dance team to perform it at the exhibition opening. Their performance was a great success and an exhibit featuring it was constructed, so that this *ute* could be displayed as one more *tivaivai*-inspired artwork in the exhibition.

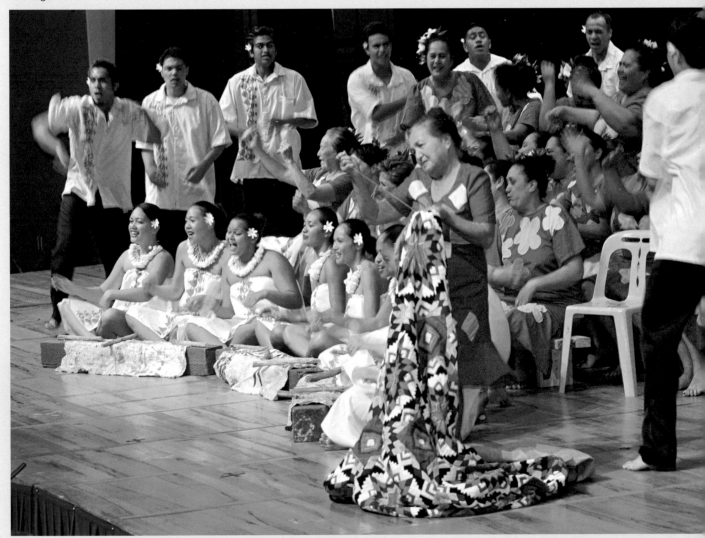

Memorable music

The Cook Islands have a rich oral tradition, and important events are still remembered through chants and performances passed on from one generation to the next. The spoken word is held in much higher esteem than the written record; the orator's skill and eloquence are often more appreciated than the precise contents of his speech. Though the essence of information remains, the stories that surround Atiu, its people and their traditions are manifold and ambivalent.

The islanders themselves did not develop a written alphabet; they trained their memorial skills instead. Chants and songs bear witness to events that were and still are important to these nomads of the sea, the rhythms helping them to memorize. Historical and practical knowledge was not necessarily available to all. Its keepers (*tumu korero*) were a selected minority, whose special talent made them eligible for such a privilege and who enjoyed the power their superior knowledge gave them (Davis 1992: 50). Even today important events are remembered in songs specifically composed for that purpose, their authors receiving the communal recognition they deserve.

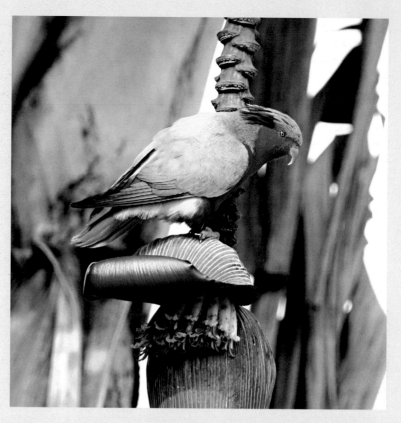

True to that tradition, in April 2007 a contest was announced to compose a song in commemoration of an important conservation project: the 'return' of the *kura* (*Rimatara lorikeet*) to our island. This colourful lorikeet was once greatly appreciated for its bright red feathers, which gave the bird its name, but which were also responsible for its extinction in nearly all of Polynesia. Today the fact that the *Rimatara lorikeet* has no natural predators on Atiu provides it with a sanctuary in which this endangered species may be able to survive unharmed.

Quite appropriately, Atiu's ancient name is recorded as 'Enua Manu' (Kloosterman 1976: 12), the literal translation of which is 'Land of the Living Creatures', because *manu* is the general name for any living thing moving on the earth or through the air (Savage 1980: 139). *Manu* can also be translated as 'bird' (Buse and Taringa 1996: 223). Today Atiu/Enua Manu likes to be known as the 'Island of Birds' again.

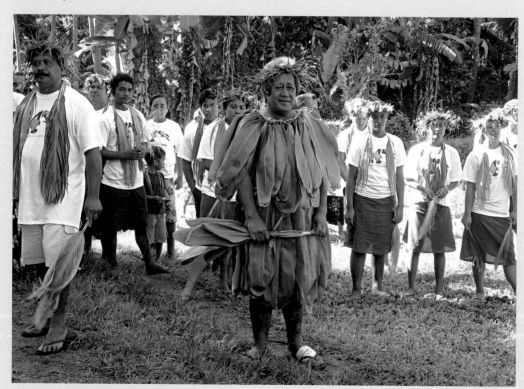

Above: *kura* (*Rimatara lorikeet*). Photograph courtesy Phil Bender.

Left: *tumu korero* (keeper of knowledge) Ina Teiotu with group of school children about to sing a chant for the release of the *kura*.

which produces a strong affinity between musical rhythm, the patterning of colours and olfaction. Encountering one conjures up the other two senses in a complex nexus of sensory perception.

Different cultures present strikingly alternative ways of 'making sense' of the world, and in their manner of doing so they frequently store and transmit ideas in ways that differ from classical theories of 'oral mentality' commonly bestowed upon cultures that do not utilize pen and paper for cultural transmission. In the Massim region of Papua New Guinea, for example, primacy is accorded to sound or 'pitch', as secret knowledge there is encoded non-verbally in the colouring and lines of canoe prows and shields – the verbalizing of such knowledge could 'offend the ears of the spirits'. For the people of the Massim, memory resides both in the belly (thus they fast when rehearsing new knowledge) and in coloured things. The Ononge of the Andaman Islands in the South Pacific, to the contrary, live in a world ordered by smell. For them, odour is the vital force of the universe and the basis of personal or social identity. Materials here carry socially effective ideas in so far as they can be subject to decay and ageing, thus making materially tangible processes that bind the visible and the invisible world together (Howes 2003).

One might conclude that the Cook Islands, famous for its floral perfumes which are produced for export, value the sense of smell over other sensory modalities in similar ways to the Ononge. Yet *tivaivai* tell otherwise, by foregrounding the intriguing relations between colours and olfactory associations, which are further supported by associations with pitch and the sound of colour terms. 'White', for example, is '*teatea*', emphasizing with its open vowels an expanding sound linked with escaping odour. The term is similar to the that for 'spirit', '*vaerua*' ('*vaī*' – to wrap; '*va'a*' – mouth); while the name for red is '*muramura*', which with its dark sounding vowels calls up associations that point to durable containment in materials that allow for no escape, such as '*tupuna*', the word for ancestors commemorated by cement grave-structures, or the coconut tree, called '*tumu nu*', whose fruit-juice is imprisoned within the coconut shell.

Colours can appear to have a factuality about them that closes off any introspection. From Locke and Hume to Moore and Russell, colour has been taken as paradigmatic instance of a simple un-analysable quality. Only in the twentieth century did a critique of the notion that colour is a property of light appear, and were colours explained by how they are perceived. Today, we conceive of colours as qualia, that is as properties rather than essences of things, and they have come to be associated with femininity, sexuality, intoxication, addiction and savagery; and with the 'stickiness' of things that cannot escape the moment. There have been a number of diverging approaches to the analysis of colour, fluctuating in popularity, perhaps the most common of which has been to understand it as a symbol. Early twentieth-century studies assumed a relation to exist between features of the language used to describe things, such as limited colour terms, and the way people think. Research by Heider and Rosch (1972) with the Dani of Papua New Guinea showed, however, that focal

colours are recognized independently of lexicon. This research prompted the investigation of basic colour terms by Berlin and Kay (1975), who argued that all languages followed a universal evolutionary pattern of colour names: from black, white and red to eleven terms in Indo-European cultures.

The universal sequence proposed by Berlin and Kay seemed to concur with ethnographic research, where there is a well-documented ritual triad of black, white and red. Yet colour is not totally described by the Munsell 'internal' colour space of hue, brightness and saturation. The way we see and experience colour is also a product of its relational quality: its relation to other colours in the perceptual field and its relation to other sensory modalities. When we consider colours' 'inherent relatedness', we see their inter-relationships are systematic, just as ordered as colours' relationships to wavelengths or degrees of temperature. This recognition of the material and relational qualities of colour was an important part of Johann Friedrich Goethe's colour theory, which had a profound influence on German Romanticism in the early nineteenth century and has undergone a revival of interest in recent times.

Colour was defined then, and again now, not as symbol but as material substance, an idea that influenced synaesthetic artists such as Kandinsky and Klee in the early twentieth century, who rediscovered relations between colour substances and sensorial capacities in their art. Using Kandinsky as inspiration for contemplating the nexus of sensory modalities that can be made visible in patterned compositions, one notices the correlation that exists on the surface of *tivaivai* between colour and tonality and rhythm. As the light colours of cut-out and appliquéd flowers appear to bounce off the usually contrasting background, tilted to one side as if in the midst of rotation across the surface, one is instantly reminded of another art-form that is prevalent in the Cook Islands, and in Eastern Polynesia as a whole: dance and song. The polyphony of colours on the *tivaivai* – dark and bright in strikingly complementary shades of red, blue, violet, orange and yellow – in compositions repeated in a punctuated rhythm, remind one of the action dance, called *kappa rima*, which is performed by men and women who move across the stage in tightly ordered formations that draw symmetrical and rotating patterns on the floor. Supported by the dark rhythm of the drumming for which the Cook Islands are famous in Polynesia, the long white skirts worn by the female dancers, made of the inner bark of lemon hibiscus tree branches or of artificial fibre, appear to the eye much like the rotating flowers on a *tivaivai*: telling a story with hands, hips and feet that complements the embroidered blossom, petals and leaves of the flower heads.

As between dance and sewing, we can easily uncover an analogical relationship between pitch in song and the sewing of coloured patches of cloth. For instance, the type of song called *imene* (the popular form of which is known as *tuki*) can be likened to a *tivaivai taorei*. It has a hymn-like composition that makes it an appropriate performance for church, and with its repetitive melody, different versions of *imene tuki* are recognized by the types of pitch amplified within the song.

The more traditional form of song in the Cook Islands is a chant-like form called *ute*. Although school children perform these in Schools Culture Festival competitions, this style of singing is still largely the prerogative of older people. The traditional *ute* was a joyful love-chant song or *imene akaepaepa* (song of praise) performed by a group of men and women in celebratory mood. Nowadays, *ute* are composed about a wide range of subjects, all concerned with topical matters such as the environment or politics. Like a *tivaivai tataura*, whose appliquéd flowers recall recent trips abroad or the visit of relations, the *ute* recalls what is currently happening in Cook Islands life. A third form of song is called *pe'e*. These are ancient historical chants, which, like a *tivaivai manu*, commemorate particular events, including brave deeds of ancestors or legendary warriors. Like a two-colour *tivaivai manu*, *pe'e* are formulaic in structure and ritualized in presentation. Their very nature is the reason few traditional *pe'e* survive. Because they were ritualized and could only be chanted by certain people at certain times (all rites which were considered heathen by the missionaries), many ancient *pe'e* fell into disuse and were consequently lost. They were once, though, the mainstay of *'eva*, the entertainment festivals which appear to have predominated in Rarotonga at the time of the arrival of the missionaries. *Pe'e* has now gathered a more general meaning referring to any chant, old or new.

While we may once have considered colour to be merely symbolic, we have now seen it to possess a material quality that testifies to the plural connections that underscore daily life, giving it a tonality and a sensory resonance that deeply impresses itself on the mind. The association between colour hue and the pitch of sound provokes an outpouring of creative articulations across different media. Together, they form a kind of glue, which is holding Cook Islands visions of self and society together. It is making the sensory modalities of social life tangible in material form that allows the *tivaivai* to recall persons and events alike, and to transmit such memories visually across generations in ways that make the *tivaivai* the most treasured possession of all.

The world in a garden

Flowers, we have seen, have the capacity to invoke family resemblances, a kind of botanical order that allows us to find connections where we might otherwise have discerned none. It may not surprise us, therefore, to find that just as *tivaivai* move from the living to those who have departed, into their graves and into the metropolises of the Pacific and beyond, so flowers move in the opposite direction, back to islands. Gardens, much like the *tivaivai* that lie hidden for most of the year, are thus emblematic of the worldly connections that forge the heartbeat of Cook Islands identity.

Visits abroad by Cook Islanders are frequently within groups, whose journeys are usually organized under the auspices of the church. It is mostly women who conduct such shorter trips, while men tend to leave for extended periods to work abroad; although this is a generalization to which one can find many exceptions. The trips are financed by selling patchwork *tivaivai*, and also involve the making of *tivaivai* to be gifted on arrival. The flowers which decorate the gifted *tivaivai* are mementos of such trips, and this can clearly also be said for the flowers which grow in the front garden of every house on the islands.

For an anthropologist interested in how people are related to one another, the flower gardens in the Cook Islands prove an enormous help, both in terms of prompting women to release information about relatives that have departed (to the land of the dead or to new shores), and by serving as a visual signifier of the social relations that drive political and economic processes on the islands. Sewing bees, groups of women with a collective bank account from which they purchase cotton fabric and who work together on the tasks of cutting and sewing *tivaivai*, also share flowers. Their gardens will be as alike as each other's houses, as they use the proceeds of selling *tivaivai* to make sure each has equivalent domestic equipment. They are not generally related genealogically or by marriage, but are usually friends, often even living in different villages, who are acquainted with each other through work or sport.

Sewing bees are of increasing economic significance in the islands, now that many able-bodied men have emigrated to work on large-scale construction sites, initially only as migrant labourers, but often staying indefinitely. Women in the Cook Islands run everything, from the schools, to the hospitals, to the airport and the shops, and the connections they have come to form under the auspices of the *tivaivai* form the vital support structure that regulates the distribution of goods and services on the islands. It is a fragile world indeed, which is increasingly reliant on the stream of money being returned by relatives living abroad, a flow drawn to the islands by the counter-movement of *tivaivai* that serves to remind the diaspora of their lasting connections, much as the flowers in the gardens remind those who remain of ways of securing the future. Stitching flowers, as much as tending to them in the soil, is thus not just a treasured pastime, but one afforded with an air of necessity and of urgency, a testimony of a material knowledge which enables women to craft a future where there may as well be none.

The flower in the Cook Islands speaks of a trading identity hidden in the light. For what we look at in the shiny colours of petals, pollen and leaves, which appear to have factual and straightforward presence, is actually the catalyst for memories that outlast the fragile existence of the flower. It is thus in spite of, and also because of, their ephemerality that flowers figure so prominently in Cook Islands social imagination and art.

Notes

1. The waxy upper part of the leaves on a coconut frond that have been separated from its vegetable underside and dried, they are known as *rito* (in Rarotonga) or *kikau* (in Atiu).
2. The Tahitian god Oro has been considered similar by some scholars (Gill 1865, vol. 2: 27) to the Cook Islands god Rongo. Oro is descended from the superior god Ta'aroa or Tangaroa, who emerged forth from darkness when earth and sky separated.

Cuttings and seeds

'Can I have a cutting of your orchid?' asks my Aitutaki friend while we are having a cup of coffee on my veranda, beside which they bloom. 'Well, if that's OK with your island's agriculture regulations,' I reply, conscious that the exchange of plants between islands may not be such a good idea, because it could also mean the introduction of plant diseases and pests. But secretly inside I am proud, as for a short time I can feel like one more Cook Islands woman.

I have absolutely no gardening skills and can rarely reciprocate when friends give me a new plant or allow me to collect a cutting from their mostly immaculate gardens. Our garden area is huge, the soil depleted by the pineapples that once grew where our house now stands. It is too hard to grow flowers here; in some areas not even the weeds want to grow. Quite frankly, I'm not the gardening kind. But I sometimes wish I knew more about the secrets of gardening, as I would be able to feel the same excitement my Atiu friends must feel when a cutting they bring home takes root and rewards them with the first flower, one that none of their friends has in their garden, and which many will come to admire and wish to be allowed to take a cutting or seedling from for their own garden.

'The anthurium cuttings I got from my aunty on Rarotonga,' my friend tells me. 'She has all sorts of flowers growing in her garden in Titikaveka [Rarotonga]. She always wins the home gardening competitions. Just recently, her photo was in the newspaper when she had won yet another reward,' she adds proudly.

I prefer to dedicate my time and creativity to cutting plants and flowers from fabric, rather than investing it in forcing plants to grow from an infertile soil. I prick my fingers with the embroidery needle, not with the thorns of some puny roses. My flowers do not wilt. They may produce neither cuttings nor seeds, but they do inspire other women (and men, perhaps) to create patterns for their own *tivaivai*, one day perhaps forming a loving gift, a cherished heirloom piece or an award winner in a national or international show.

Embroidered orchids (detail - work in progress, Andrea Eimke).

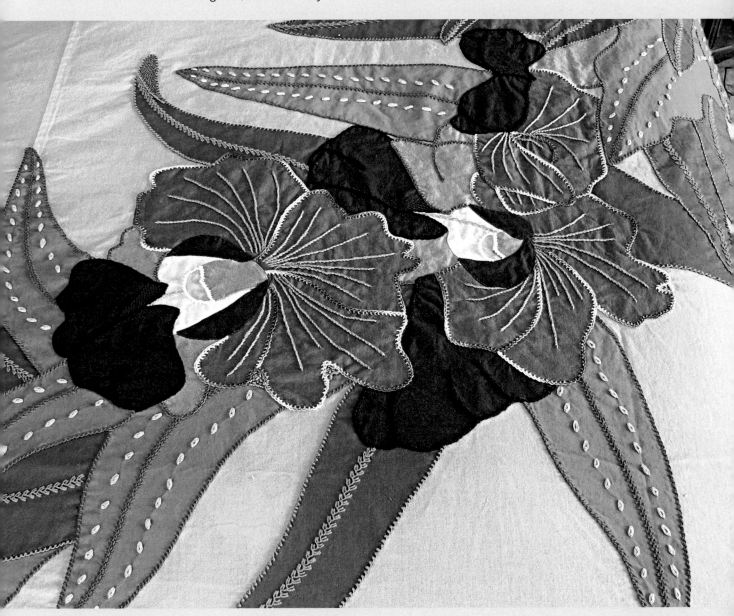

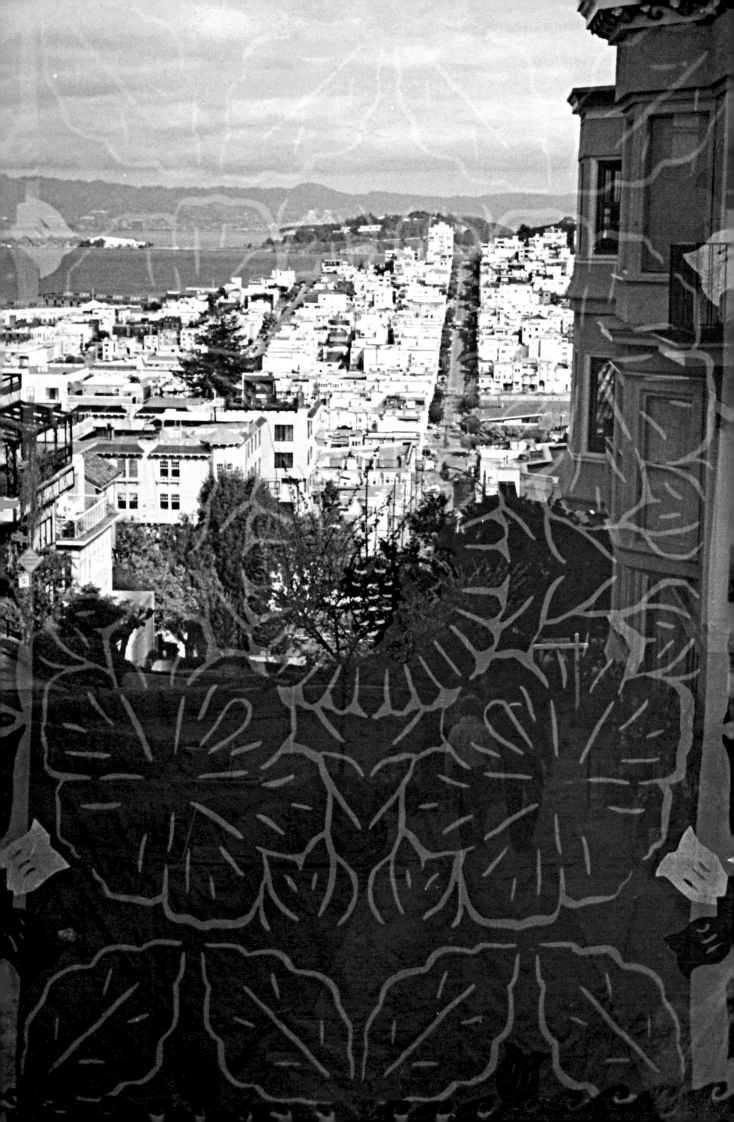

Chapter 5

From social fabric to textile art

Thus far, we have discussed *tivaivai* as exerting a number of kinds of significant social agency. They have been shown to have assisted in the historical reorganization of society, from a system based on the transmission of divine power through the male line of descent to one in which power is channelled through the female line. It has also become clear that they have been vital to the transmission of biographical knowledge and connections across expansive and interlocking fields of relations, which upon independence came to encompass the entire British Empire. This capacity is in part down to their salience to social memory and the mapping of biographies. Finally, *tivaivai* are supportive both of a floral aesthetics which sustains an appreciation of light as material substance, and of a disposition to value practices that harness and transform light into immaterial substances such as odour, or into abstract concepts such as life or power. *Tivaivai* in the Cook Islands are clearly more than well-made objects for home decoration, or the product of a 'hobby' undertaken in the quiet hours of the day. Rather, they give us profound insights into the way Cook Islanders 'understand' the material and visual world and each other, and the manner in which they articulate this as a way of sharing attitudes to belonging, which have become of increasing importance in the diaspora, where most Cook Islanders live today for at least part of their lives.

The previous chapters have shown that *tivaivai* signal a paradigm shift in the way connections are imagined and calculated, in that immaterial connections grounded in genealogy and cosmology have begun to be supplanted by material connections maintained through the gifting of *tivaivai*. People's response to *tivaivai*, however, has changed over time, as these new material connections have grown in their importance to the sustaining of localized political economies. The agency that has long been attributed to patched cloth created the fertile ground for these fabrics, saturated with social and cultural importance, to take to the stage of 'contemporary art', where they have come to figure prominently in conveying ideas about place, culture, history and politics. From the earliest public outings, such as at the New Zealand Maori Women's Welfare league in 1975 (Metge 1976), to the first exhibition as artworks in their own right at the *Te Moemoea no Iotefa*/The Dream of Joseph exhibition in 1989/90 (Thomas 1996a), and the most recent public acknowledgement of the Pacific Textiles Project featuring mats and fabrics at the 5th Asia-Pacific Triennial of Contemporary Art in Queensland in 2006, *tivaivai* have invited us to consider how apparently humble and utilitarian things constitute models of exchange, genealogy and politics that map time and history through the mode of *art* styles, genres and practices. This chapter will discuss the take-up of *tivaivai* by the Pacific contemporary art scene and the difference that it is making there to the perception and study of textile arts today.

To understand what enables *tivaivai* to speak to a wider Pacific audience we must recall that *tivaivai* was a pan-Polynesian phenomenon at the outset, as variations of patterns in fabric are also sewn in Hawaii and Tahiti, where the art of stitching is known as *tifaifai* (Pogglioli 1988; Hammond 1986). We have already explored in previous chapters how although each and every patched fabric is exchanged at ceremonies marking key local and familial events, recalling in their patterns histories of their exchange and the lasting connections they connote, the irony is that the patterns themselves are not local in origin, but allude to texts that are global in reach, such as pattern and stitch books and magazines. In binding the local with the global flow of culture, *tivaivai* are the ultimate agents of modernity, activating a space for the local reception of global processes and cultural forms in which the foreign can be neutralized or demarcated in diverse and unpredictable ways. Neither quite local, nor quite global, nor indeed the two blended too smoothly, the visual dynamics of *tivaivai* illustrate what Nicholas Thomas has called 'the need [for] a more complex and nuanced vision of the complexity of the postcolonial world', in which cultures are neither purely exotic, nor derivative expressions of the West, nor presentable as 'hybrids' (Thomas 2006: 29; Thomas 1996b). We shall see how *tivaivai*'s entrance onto the world stage of art shows and gallery exhibits complicates the blanket category of textile arts and forces us to think about their relevance as a local response to global phenomena.

Opposite: Tuatai Koronui's *tivaivai manu* in San Francisco, 1992.

Tivaivai as an art form

To the envy of quilters, *tivaivai* made it into art galleries in New Zealand earlier than their, seemingly related, padded counterparts. One of the first major exhibitions of *tivaivai* took place at the Auckland New City Art Gallery in 1995. Although not necessarily the first time *tivaivai* entered a gallery, this exhibition was pivotal in changing attitudes towards women's artworks in fabric. Visitors were prompted to engage with *tivaivai* as an art form, rather than a craft. This perception, however, raises new questions that overthrow long-standing assumptions regarding the relevance of textile arts to the study of contemporary art, which until then had been seen merely in purely cultural terms.

Most of us take for granted that art presents us with key indicators of cultural diversity and alterity, as we tend to assume that perceptual experiences are culturally coded in ways which open them to sociological as well as art historical analysis (cf. Rampley 2005). Thus we assume that the differences in pattern that single out Cook Islands *tivaivai* override the significance of the commonalities, which its patterns share with those of neighbouring cultural groups. However, this certainty in the attribution of style to distinct cultures, which has long prevailed over the study of the visual within culture, is fast disappearing. For example, we now recognize as a fact of modern times that new media and new technologies enable people to consume styles across a variety of media and to appropriate them as their own quite independently of each other or of other aspects of culture. Taking this movement of pattern and image to the Cook Islands we can see that, in much the same way that typical *tivaivai* patterns are at home in the craft magazines of Euro-America, Cook Islands-, Tahitian- and Hawaiian-style quilt patterns are now made available to manufacturing companies in cheap-labour countries such as the Philippines or China, which produce seemingly genuine patchwork for sale in souvenir shops. *Tivaivai* patterns are also used to decorate bed linen, curtains and other decorative accessories that are for sale in New Zealand. This mobility of pattern, with its many uneven and diverse local receptions, has led some to search for the dynamics underlying the creation of world art, while others examine the propensities of style itself, exploring how it is that we respond to and transmit images as recognizable and memorable patterns (Friedberg 1989; Küchler and Were 2005). Spurred on by a trust in the technical capacity to replicate a pattern, most writing on culture on the move has been preoccupied with the question of how a product is localized, how meaning is grafted, in processes of consumption, onto things that are in themselves considered insignificant, non-resistant and vacuous (Appadurai 1986).

Tivaivai's entrance into the gallery puts our received assumptions about the appropriate approach for textile arts into a fresh perspective, as it draws our attention not just to local processes of consumption, but to the materiality and the technique of patching, which resists being restricted to a localized culture. Patched fabric, in fact, tells us about the take-up of fabric and sewing technology, and the West's entanglement in this story, not least because of the material and technical affordance (perceived action possibilities) of fabric, which swept the Pacific as a 'new material and new technology' at a time in the nineteenth century when Euro-American interests were similarly positioned: searching the colonies for new materials that could invigorate manufacture and sustain the newly fashioned plantation economy (Schiebinger and Swan 2005; Smith and Findlen 2002). We have shown how the take-up of this new material, and of its allied technique of the sewing of patchwork, supported a new vision of society in the Cook Islands; here we will ask not what motivated this take-up or what in turn it unleashed, but how it may enable us to ask new questions about textile arts in the Pacific.

The Cook Islands Tivaivai Association

Whenever I have the chance to spend some time on Rarotonga, I like to meet with friends involved in the *tivaivai* world. One of them, Sonya Kamana, had worked for many years at the tourism office. Both of us remembered the *tivaivai* exhibitions organized on Rarotonga by the Tourist Authority. Both of us also noticed that fewer women were now making *tivaivai*. What could be done to encourage their production, to promote traditional textile art nationally and internationally, and to attract visitors to the islands who would appreciate this unique art form? Perhaps, we thought, it would be a good idea to organize an international *tivaivai* festival. The festival was to be organized for a time of the year when tourism was low and airfares were on special offer. Hotels would be only too happy to offer good rates, so long as their rooms were filled. This would suit our visitors' budgets, as I knew from experience that textile people were not necessarily affluent. We could invite some overseas quilt tutors to give workshops in their special techniques, and some of our *mamas* could be invited to teach our visitors the intricacies of *tivaivai* making, establishing an exchange of technical knowledge. Both local *tivaivai* makers and overseas quilters would be invited to participate in exhibitions and competitions. There could be smaller *tivaivai* shows held on the outer islands and our local airline could organize day trips there with interested visitors, so that the whole country would benefit from the festival. Those women who were interested could make *tivaivai* for sale, local artisans would offer their carvings, baskets, hats and mats. The ideas kept pouring from our heads. In our minds' eyes we saw colourful textile works on display, not only in the auditorium and the museums, but also in banks, hotels, churches and other public buildings. We could almost hear the *mamas'* happy singing and smell the fragrance of their flowers. It would be a superb event and the whole country would benefit.

At one level, such new questions can be seen to resonate with recent literature on art that draws our attention to a way of inquiring into what an artwork does that reaches beyond culture: exploring the nexus of generic relations that may exist between persons and things. Art historian Barbara Maria Stafford has recently (2007) built upon this tradition of scholarship, and has begun explorations of neuroscience and consciousness that may open our eyes to the complex echoing between art's imagery and the mind. Stafford's work on the relation between art and consciousness builds upon the recent scientific discovery of so-called 'mirror neurons', which we now know search out patterns in the material and visual world that are analogous to those formed by the thinking mind. Artworks, according to Stafford, can be seen to possess agency in inciting our innermost thoughts and emotions into action and in pointing up to us, like a mirror, the kind of relations between people and things that are important in our lives. No longer are artworks exemplars or reflections of meanings produced elsewhere. Instead, we return to artworks themselves as *sui generis* 'meanings' and can reflect upon what they do in society.

That artworks have agency in presenting analogical ways of thinking about the world around us was also an argument put forward by the anthropologist Alfred Gell. His posthumously published work *Art and Agency* (1998) broke with a long-standing tradition of narrowing the anthropological and art-historical enquiry to the sociological analysis of artworks that are produced exclusively for an art market. Instead of assuming that artefacts and performances are responded to by values that are socially and culturally situated, he proposed to search for the conditions that allow artefacts, or for that matter performances, to direct attention in a manner typically attributed to artworks. This concept was arguably already present in the work of Matisse when he drew inspiration from Tahitian *tifaifai* and nature to create

works that fashion a new visual environment. Like Stafford, Gell finds the answer to what patterns do in the indexical, and thus logical, relation between entities that are usually held distinct; a relation which finds its mimetic expression in style, whose psychological saliency rests on drawing attention to the synergy that binds artworks to one another.

Heralded by this vision of art is a new sensitivity to the material and visual properties of artworks, which have been much neglected despite a surge of anthropologically orientated studies. Gone is the time when an artwork could be interpreted as reflecting a meaning located in the social context of its production and consumption; instead of looking beyond the artwork to uncover its meaning, we are now asked to look at the material and visual relations it brings to our attention. As we identify the performative, textured and material dimensions of an artwork, we ask what it does in society, rather than what it means, and probe towards the material and visual properties that make manifest immaterial and intangible connections that people think with when negotiating all sorts of relationships. (continues page 100)

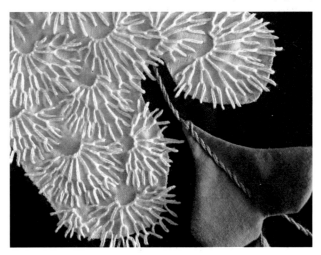

Details of embroidered wall-*tivaivai* 'Coral Garden'.

We spoke with friends and soon the C.I. Tivaivai Association was founded, registered and incorporated. Our next months were filled with finding tutors, sending off advertisements to textile magazines in many other countries, establishing connections with the rest of the textile world, which, so we were convinced, would love to come to our islands to celebrate with us. And our expectations were right. The response was great; many replied and even booked their participation. However we had made one severe mistake. Since the country's economy was so depressed, we did not want to approach our local businesses for funding. We knew the same companies were always asked, and that at present they had reached the limit of their possibilities. We received funding for a feasibility study that showed that, if the first festival was funded, it would generate enough funds to repeat the festival in regular intervals. From international quilters' magazines we compiled documentation showing that similar events for quilters in other countries had attracted tens

of thousands of visitors and that quilting in the United States, for instance, is a multi-million-dollar industry. Unfortunately, the Australian and New Zealand businesses and funding agencies that we contacted did not want to know. When we realized that we would not be able to meet our aims, and that the money that Jürgen (my husband) and I were advancing privately was starting to reach proportions our budget could not withstand, we had a meeting with the management committee. We decided to call the festival off.

My next weeks were spent sending out the cancellation letters to tutors, to potential participants, to magazines recalling advertisements, and the like. It was a sad time. Extremely comforting for me were the sympathy, warmth and encouragement with which most people responded. The day after all the letters had been sent, we received information from one of the major potential sponsors that we had approached, offering his assistance. Too late!

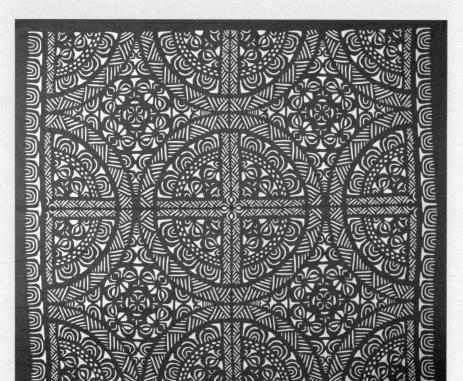

'Tifaifai pa'oti' (snowflake-style *tivaivai*) designed and sewn by Emma Tamarii, Tahiti.

Islanders visited the exhibition, too. I was particularly pleased by the interest of one fourteen-year-old, with whom I spent nearly an hour looking at the *tivaivai* and answering her many questions. This young girl's interest showed me that our hope that *tivaivai* making can be kept alive is not in vain, that there are young ones who perhaps will continue this wonderful tradition.

Studying Cook Islands *tivaivai as an art form*

Fred Hagstrom of Carleton College in Northfield (Minnesota, USA) had been to Auckland and visited the *tivaivai* exhibition 'Patterns of Paradise' at the New Auckland City Art Gallery in 1995. As a result, his college established biennial study trips for its students to New Zealand and Australia that included a stop on Rarotonga. Amongst other activities, the students spent time with a number of Rarotongan *ta'unga* such as Vereara Maeva and Tokerau Munroe, and *mamas* of their sewing groups, learning about the designs and techniques of *tivaivai* making. The young visitors must have felt sorry for their fellow students back home, because soon the idea was born to organize a *tivaivai* exhibition at the art gallery of their college. As international liaison officer for the Tivaivai Association I was requested to curate this exhibition, displayed in September 2000. The *tivaivai* makers on Rarotonga and Atiu were proud to lend us their works for an exhibition overseas, especially since they had read about the visiting student group and their study programme, or had even met them in person during the workshop. I had planned that year to travel to the United States to visit a quilter friend of mine. I organized my trip so that I would be able to attend the opening of the exhibition. The college kindly provided me with accommodation for my short stay. *Tivaivai* displayed in a gallery never fail to excite me. Once again, the beauty of the women's works, the vibrancy of their colours, and the intricacy of their stitching enchanted me. Laurel Bradley (then art director of the college gallery) and her team had displayed the works to their utmost advantage. They had even been able to find a bed whose four posts were adorned with carved pineapples to display an unusual *tivaivai taorei* whose motif was pineapples (see photograph page 20). Tutors and students made me feel very welcome and showed much interest in the slide presentation and in receiving as much information as I was able to give them. I was told repeatedly that they much appreciated the fact that these colourful ambassadors of an exotic foreign world had come to those who had not been able to visit the Cooks themselves.

After the 1999 success, the CITA decided to hold a *tivaivai* exhibition at the National Museum every year from mid October

'Reflections'

But our failure did not discourage us from the other aims of the Tivaivai Association. If we could not have an international festival in 1999, we should at least have an exhibition! Local businesses helped with some funding for the installation of the exhibits, the invitations, etc. Rarotongan school children drew the posters to be hung in each shop window, each hotel and everywhere else that mattered. Our president managed to convince Rarotonga's leading *tivaivai* makers and owners to lend us their heirloom pieces for the Association's first exhibition. I did the same on Atiu. Together we compiled an impressive display. The ages of the *tivaivai* spanned over sixty years; and these textiles had been kept in a humid tropical country, where mould and insect damage are an everyday threat. If one had added up all the hours that it must have taken to make each exhibit, and counted all the stitches contained in the beautiful works, they would have reached far beyond billions! We decided on 'Reflections' as the title, as the works reflected the creativity and skills of Cook Islands women and the exhibition was looking back in time, showing the development of technique and designs from the beginning. It was a breathtaking sight.

We had a grand opening, which received enthusiastic feedback from our invited guests, who had all come despite the pouring rain. The following two weeks I spent at the museum, studying the works. This also gave me the opportunity to talk to visitors, giving them information about our exhibition and receiving their positive comments. Cook Islands Television, who had not been able to attend the opening ceremony because of other commitments and had instead come the day before, came again to interview a number of foreign visitors. The evening news reported that the beauty of the exhibits had overwhelmed most of them. Many had been surprised to discover that the Cook Islands had their own kind of quilt. Of course many Cook

to mid November. We wanted this show to become as well known as 'Tiare Week' (flower week), and to attract visitors to our islands who would be interested in the Cook Islands' unique textile art.

Overseas adventures
A tough lesson in Spain

When I read that an international quilt exhibition would be held in 2002 in Barcelona, I felt that participation in an internationally known textile art fair could be an excellent opportunity for Cook Islands *tivaivai* artists to make their art known beyond the boundaries of our island nation. We discussed the matter, all agreed and I approached the organizers. They replied with an invitation to CITA for participation as exhibitors and to the Fibre Arts Studio as vendors. Soon we were busy trying to raise funds for this special event. Local businesses were generous with their financial help and we were able to secure a grant from the European Community to cover shipping costs and travel expenses for both the Tivaivai Association and the Studio.

Organizing participation in an exhibition so far away from home is quite a challenge, much more so when living on a remote island, where the benefits of email had just been established, and where both telephone and email use were still exorbitantly expensive, making access to website information impossible. The biggest difficulty lay in obtaining information on customs regulations in Spain regarding the documentation our exhibits would need. We were neither able to obtain the requested information in the Cooks or New Zealand, nor from international freight-forwarding companies in Europe. The Cook Islands seemed just too remote and unknown: the forms we were asked to produce were not obtainable, and the benefits of the trade agreement which the Cooks had just signed with the European Union would not come into effect until a year after our exhibition, hence the new forms were not even printed yet!

We shipped the exhibits anyway, hoping that things would work out all right. For the sales booth that the Fibre Arts Studio had rented I took a collection of works in my suitcase and just had the list of items stamped by customs when leaving Rarotonga. As they were supposed to be for sale, I declared them upon arrival and even though I did not carry the correct documentation, the customs officer was prepared to let me pass after I had paid the duty.

The setting up would begin the day after Easter, in Spain one of the most important holidays, and a day before the fair was going to be inaugurated.

'Ah-tissue' by Krick Barraud.

Sonya and I arrived at the Barcelona fairgrounds early. Much to my dismay, nobody had the faintest idea where our exhibits could be: they had obviously not arrived! After a long investigation we finally discovered they were still held at the airline's airport office. We were told that we would need to deposit a bond of 67,000 Euros to get them out of customs for display! We were devastated. Even though our 'quilt angel' Monique knew every possible office and contact, we had no other choice but to produce the bond if the exhibits were to be displayed.

While Monique kept trying other avenues, Sonya and I set up the Fibre Arts Studio's booth. The *tivaivai* and wall-hangings I had brought with me soon created a happy South Pacific island atmosphere, even though on the inside we were not happy at all! The opening day was the next day and the exhibits were in customs. Early in the morning the organizers asked us to their office, and we were introduced to the president of the Spanish Quilters Guild. She and her wonderful members had borrowed the 67,000 Euros and would present it to customs as bond to release our exhibits!

The joy lasted but a short while. It was Friday morning. We went straight to the customs office, where we were told that they were very sorry, but their office would close at 1 p.m. and there was no way to get the documentation completed to release the exhibits before Monday – though the fair finished on Sunday! No pleading helped, the fifteen heirloom pieces the Cook Islands

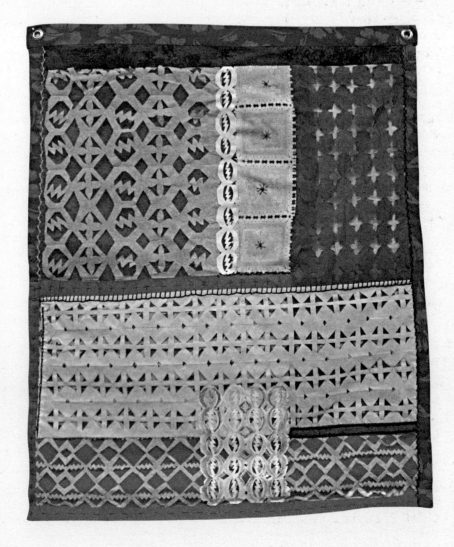

mamas had so kindly lent us for display were never shown to the Spanish public, nor could the generosity of the Spanish quilters help our cause.

We were not the only country without exhibits. The Turkish quilters also had problems, but though late, their quilts finally did turn up and could be hung. And we were not the only ones with problems regarding customs regulations. Vendors from non-European-Union member countries were not allowed to sell their goods. We were one of those. Luckily, we were at least permitted to display the Fibre Arts Studio's few *tivaivai* and I had photographs of the exhibits and information that we could show the visitors, of whom we received an endless stream. All of them were hugely interested in our small island nation and its unique textile art. On our feet for twelve hours over each of the three days, we told and repeated our sad story until we had no words left. Everyone commiserated with us and admired the few works that we were able to display, regretting that they had been deprived of the whole show. At least that way we were able to introduce our islands to the world of textile art and communicate that we had an art form all of our own. As ambassadors of the Cook Islands we had certainly been successful. Though our island country's location is already impressively exotic to Europeans, our misfortune helped them to remember our unique textile art even more.

We may not have returned home triumphantly, but we were not defeated either. As it turned out, that was the right attitude to have. A few months after our return I received an email from France. The news of our problems had reached the organizers of the European Patchwork Meeting in Sainte-Marie-aux-Mines. They invited the CITA to participate in their international event in 2005 as guest of honour.

Tivaivai to *Tauranga*

Tauranga, in New Zealand's Bay of Plenty region, has become well known for its biennial 'Garden and Artfest'. Showcasing over a hundred exquisite gardens in the region, and as many local artists' works, it is all about flowers and art, so what could be more appropriate to their programme than Cook Islands *tivaivai*? In 2004 the CITA was asked to contribute to a one-week *tivaivai* exhibition. With enormous efficiency, the organizers had secured some fifty *tivaivai* from Cook Islanders resident in New Zealand, to which our association was able to add another thirty of the Cooks' best works. The venue was a huge old packing shed, whose grey corrugated-iron walls and raw beams contrasted beautifully with the colourful exhibits pegged on 'washing lines' with old-fashioned wooden clothes pegs.

CITA's president, then no longer resident on Rarotonga but in Wellington, was able to come to Tauranga for the inauguration. The French Ambassador to New Zealand and the South Pacific, who had just commenced his term in office, also accepted our invitation to attend the opening function. The local Bernina (sewing-machine manufacturer) representative had invited me and one of my employees to conduct *tivaivai*-making classes at their Tauranga shop. A great many members of the Cook Islands community in Tokoroa, who had lent most of the New Zealand exhibits, were also present. Their prayer and pride and the volume of their voices filled the enormous hall with a vibrancy that

touched everyone's hearts, as if the bold colours and shapes of the eighty exhibits had become pure beauty reflected in sound!

The exhibition was a huge success, partly thanks to the authentic atmosphere the Cook Islands *mamas* created while they demonstrated *tivaivai* making and entertained the countless visitors with their stories and happy singing. Many visitors had not previously heard of Cook Islands *tivaivai*, and kept returning to the exhibition for another look. I heard stories of women and men bursting into tears while studying the exquisite workmanship and detail displayed in such abundance.

The exhibition did not just have a strong impact on New Zealand visitors. When the exhibits were finally packed and we sat together, Tauranga 'Garden and Artfest' trustees, Cook Islands exhibitors and helpers, one woman got up to speak her mind:

> People been in, not once, not twice – people just been coming back, you know. They've been saying they just loved the show; they've been 'round all the gardens and they just kept saying, you know, 'it's so amazing!' I was born on Aitutaki and sort of brought up amongst these *tivaivai* and the language. When you come to New Zealand, like you weren't allowed to speak the language at home. Things like this – it's actually – you realize how much you have missed…
> ['Don't leave it too late!' interrupts another woman.]
> …but we have to get our *tivaivai* [cut for a new Tauranga sewing group] finished!

For me, this remark was a key experience. I also sensed what she had not said. I began to feel the need to reflect on the complexity of the culture of these economic migrants, split between several nations and scattered across enormous distances. Will they ever achieve a balance between what they have lost and gained? Their feeling of separation from, and nostalgia for, a distant home and a way of life that was exclusively theirs is so noticeable; their efforts to fit in and adapt to an alien system and a tradition they did not grow up with are so desperately difficult; the task of creating an identity that can be acceptable as exclusively theirs, regardless of the cultural and social environment they come from and live in, is so challenging. Perhaps *tivaivai* can help to bridge the geographical, social, generational and educational distances that today lie between the people of a proud island nation.

The French connection

The French ambassador was most impressed with the Tauranga *tivaivai* exhibition. Our request for financial sponsorship from the French Embassy for participation in the 11th European Patchwork Meeting in Sainte-Marie-aux-Mines in 2005 was granted. We were able to borrow the very same *tivaivai* we had taken to Spain. This time we were able to learn from our unfortunate experience three years previously. We knew exactly what to do, to the extent that we were now able to inform the organizers of the requirements. With an iron will to succeed, we were determined to overcome the minor hurdles that challenged our participation in this event, and we did. As an added bonus, a group of Cook Islands *mamas* who had booked a trip to France, and had earlier joined our association as new members, was able to include a detour to Alsace in their schedule.

The venue in the romantic town of Sainte-Marie-aux-Mines was an Art Nouveau theatre. We could not have wished for a better setting for our *tivaivai*! The most precious piecework *tivaivai*, one of them made in the 1940s, were prominently

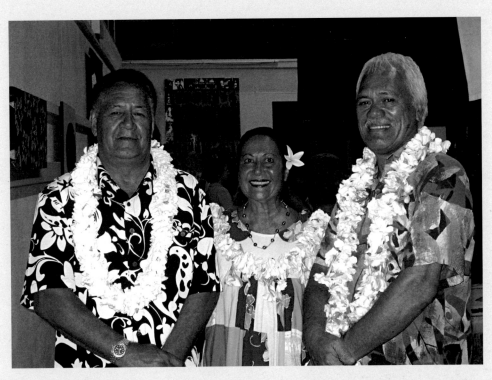

At the opening of the CITA's 9th annual *tivaivai* exhibition (from left to right: Sir Frederick Goodwin, Queen's Representative; Te Tika Mataiapo Dorice Reid, president of the CITA; Jim Marurai, Prime Minister of the Cook Islands).

displayed on stage and provided the most appropriate backdrop for one of the *mamas* who happened to have taken a *tivaivai taorei* in progress with her. She was able to practise her newly acquired French-language skills while explaining to countless quilters the secrets of Cook Islands *tivaivai* making.

Together with my 93-year-old father, I had prepared several video documentaries showing *tivaivai* exhibitions on the island of Atiu, interviews with a number of *tivaivai* makers and the manufacture of a very special *tivaivai* I had designed for a show. These were played in the foyer and attracted a huge number of viewers. Brochures and informative material about the Cook Islands were laid out on a large table, manned by a representative of Tourism Cook Islands and the husband of one of our Cook Islands *mamas*.

The stream of people never ceased. A French couple told us how moved they were to see these colourful textiles, so similar to those of Tahiti, the home country of their adopted daughter. The French Ambassador and his wife used a special holiday to honour the exhibition with their visit. Often we heard the praise that our *tivaivai* had been the highlight of that year's Patchwork Meeting in Alsace.

Competing with cousins

This success left such a positive impression that we were offered more financial support from France for our next project in 2006. Following the Ambassador's suggestion, I had made contact with Te Api Nui O Te Tifaifai, an association similar to ours dedicated to the art of *tifaifai* in French Polynesia. It was agreed to invite and host ten delegates of their association to exhibit samples of their works next to Cook Islands ones during the 2006 annual *tivaivai* exhibition at the National Museum on Rarotonga. We were also going to have our first ever *tivaivai* competition, for which we received sponsorship from the country's business community. The result was a formidable exhibition, and an exchange of traditional knowledge from which both groups and the general public benefited.

The opening ceremony on the evening of 23 October 2006 was attended by over fifty invited guests, amongst them various personalities from the council of chiefs, government and Rarotonga's business community. The winning entries were announced and prizes were presented. The response was again extremely positive, and our guests praised CITA's initiative to invite textile artists from our sister country to show their works alongside our own and for creating an opportunity to enrich each other's creativity. This joint exhibition brought the Tivaivai Association one step closer to its original aim of an international *tivaivai* festival.

The art of tivaivai as inspiration

By 2007 we felt that the annual exhibitions had displayed nearly all the *tivaivai* kept under lock and key in local glory boxes. We now hoped that using *tivaivai* as an inspiration for artists who mostly work in other mediums would provide an exciting challenge for the Cook Islands artist community. We planned to see the art of *tivaivai* celebrated in as broad a range of media and approaches as possible. In the end twenty-five artists and groups, of all ages, responded to our invitation and exhibited forty-six artworks of amazing variety, all inspired by *tivaivai*.

On 22 October, the Cook Islands Tivaivai Association's annual exhibition at Rarotonga's National Museum opened its doors for the ninth time to a selected audience of Cook Islands personalities. We felt very honoured to count the Queen's Representative Sir Frederick Goodwin and his wife and Prime Minister Jim Marurai amongst the many invited guests.

The exhibition displayed only few traditional *tivaivai*. To the great delight of its six creators, the multimedia *tivaivai* of the Creative Centre, a day-centre for those with special needs, was many viewers' favourite.

An *ute* dance about *tivaivai* had won the first prize in this year's constitution celebrations, so we invited the Takitumu district's dance team to perform it at the opening (see page 87). This was a special asset to the opening function, and a display was made so that the story of this *ute* remained on view during the entire exhibition period.

The unusual exhibition received an enormously positive response from artists and viewers alike, many of them hoping that similar events will be repeated in the future.

It is perhaps a symptom of the neglect of the material and technical dimensions of artworks that studies of textile art tend to be lacking in theoretical acuity, confining themselves largely to the local expression of coverings whose nature is rarely questioned. Patterns in fabric are widely shared and globally distributed, yet we understand very little about why some textile arts, such as wax-print cloth produced in Holland for Indonesia, yet taken up in West Africa, take root in certain societies and not in others. In fact, patterns in textile arts have been given the least attention of all art media, despite the ground-breaking work of Gottfried Semper (2004) in *Style in the Technical and Tectonic Arts: Practical Aesthetics*, originally published in two volumes in German in 1861 and 1863, which defined textile patterns as the springboard for innovations in the structural design of architecture and sculpture.

Standing out amongst the scanty research on textile arts is the work of Rozsika Parker whose book *The Subversive Stitch: Embroidery and the Making of the Feminine* (1981) calls for a re-examination of the nineteenth-century sublimation of textile arts in Britain to industrial production, which progressed in parallel with its confinement to the Victorian parlour. Parker shows how textile arts, such as the thread-based embroidery, lace or knitwear that we have since come to regard as decorative and as suitable only for 'pastimes', that is for time not spent on work, were prized, treasured and exchanged in a manner that complemented, in its inalienability and encapsulation in the home, the rapid and expanding circulation of manufactured cloth (Mukerji 1983). While twentieth-century Britain saw the rise of industrial fashion and its acceptance as art, stitch-based textile arts were forced underground, where they continued to forge powerful connections between women across household and even national boundaries. Such developments, which separated textile arts from the world of cloth and fashion in Britain, had an inevitable bearing upon the unfolding of patchwork in Polynesia, supporting its acceptance by mission and colonial powers as suitable to the creation of a thriving economy of the home. That *tivaivai* managed to explode the boundaries of the home and to sustain the material connections of a thriving transnational community was quite simply the result of the misrecognition of its pattern as inconsequential and ineffective.

Returning to the art of *tivaivai* and its significance in shifting our understanding of textile arts to a focus on the roles of the material and technical in take-up and transmission, we notice that such questions are inseparable from the patterns that allow us to recognize and 'place' a *tivaivai* in time and space. This volume has already explained much about *tivaivai* patterns, at least from a Cook Island perspective; we have learned, for example, that they fall into three main types of construction and modes of assemblage; and we will have begun to appreciate the fact that their floral imagery is depicted in precise and standardized symmetries. Now we continue our examination of the properties of these patterns, which connect Cook Islands patchwork with those made in other Maori-speaking cultural groups and which allow *tivaivai* to enchant whoever has the privilege to receive

one as a gift. The distinctive contribution of the art of *tivaivai* to textile arts will become clear.

Patterns that connect

Pattern has been largely examined in relation to the kind of symmetry it produces, although little work has been done to move beyond a general description of the symmetry to an investigation of logical and mathematical rules of construction. Dorothy Washburn and David Crowe have done groundbreaking work in the analysis of symmetry in *Symmetries of Culture* (1988) and *Symmetry Comes of Age* (2004), which not only give us a glimpse of the astonishing variety of symmetrical designs produced across different cultures, but also of the reoccurrence of symmetrical configurations the world over. Knowing that such variations and repetitions in symmetrical patterns exist, however, does not help us to understand what difference a particular symmetry makes to culture and society, nor how it is distributed and transmitted in ways that are surprisingly resistant to change.

The simple fact of symmetry, that care was taken to be mathematically precise and attentive to the appearance of the pattern, does not tell us much about what such pattern is seen to do in the society in which it is produced. For example, we know that New Zealand Maori design features both bilateral symmetry and asymmetry, yet despite many attempts by anthropologists to relate this dualism to features of Maori social organization, it may, as Alfred Gell argued (1998: 160), signify nothing more than an expression of relative disregard for the mathematical exactitude of symmetry.

There has been a tradition of scholarship that sees the symmetry of patterns to be an expression of structures that amplify underlying cultural norms and structural features of social organization. The anthropologist Adrienne Kaeppler, for example, argued that in the island kingdom of Tonga structurally embedded hierarchy between commoners and aristocracy is expressed in the aesthetics of an underlying structural opposition of foreground and background which resonates across song, dance and visual arts (Kaeppler 1978). Another anthropologist, Allen Hanson, similarly proposed to have uncovered within New Zealand Maori society 'homologous relations between artistic forms and other structures or patterns of culture' (Hanson 1983: 79). Correctly identifying a preoccupation with the creation of manifold symmetries through rotation or superimposition of motifs and geometries in New Zealand Maori art, he proceeds to relate this to the New Zealand Maori cultural emphasis on overlapping sequences of escalating reciprocal competitive exchanges. While interesting, such explanations, which draw on criteria external to the artwork itself to account for why patterns look the way they do, do not help us to understand how such patterns are made, or how they may be translated across different media and transmitted in society. Rather than perceiving the symmetry of pattern to be a symptom of underlying, and already existing, patterns of social relations, there is nothing in principle which prevents us from assuming the opposite; namely that patterns have agency built into them due to the fact that they are made for retention, and are thus capable of giving structure to perception in society (cf. Were in press).

Folklore studies, which are popular in America, have allowed scholars working within the field of mathematics and computing to approach the same material with questions aimed at understanding the cognitive operations required to sustain the expression of patterns and their symmetries over time. The work of ethno-mathematicians, in particular, has shown us that shapes composed of lines and numerical sequences are not only expressive of the universal rules of measurement and logic, but that patterns may also be seen to give expression to effective ways of securing the condensation and schematization of complex information, so that it can be transmitted in society. Patterns are also thus shown to connect different areas of life, which are experienced as distinct, as a result of the way they schematize complex knowledge and ideas. As they provide a way of 'seeing connections' that are not usually amplified in everyday life, patterns enable persons to 'navigate' society effectively, to predict outcomes of decisions and to devise strategies for planning the future (Ascher 2002; Küchler and Were 2005).

Patterns in fact animate the mind, as the mind is encouraged to retrace the movements that have connected the lines so as to create distinct patterns. According to Washburn and Crowe (1988), patterns are the result of variations on four rigid motions in the plane:

1. reflection;
2. translation;
3. rotation;
4. glide reflection.

In the words of Alfred Gell (1996), patterns made by motions of hand and eye in the plane work like 'mind traps', at the same time arresting and provoking our thought. There are numerous examples across the world that show how patterns that attract our attention also serve apotropaic purposes, warding off spirits or shielding contents. In India, threshold designs painted on the doorsteps of houses are thought to trap evil spirits, while locking positive powers of animation inside the house. A similar duality of function, warding off and holding in, is bestowed upon patterns tattooed on the skin in Polynesia. Whole-body tattooing, including facial tattooing, was practised in pre-colonial times in certain parts of Eastern Polynesia, where it served to insulate and deconsecrate a warrior, providing protection through the disorienting effect of the pattern, but also holding divine power (*mana*), held to be dangerous to ordinary persons, within his body (Thomas 1995). As the restoration of sanctity required the removal of the tattoo, in some parts of Polynesia chiefs of the highest rank were never tattooed, but wore wraps of barkcloth made from the bark of the paper mulberry tree, which was decorated with similarly disorientating patterns created from stamps made from bamboo slivers, drawn freehand, or carved into rectangular '*kupesi*' (Samoan spelling) design boards or wooden beaters.

Pattern is rarely random, but carries information as order and as systematic relation between parts and wholes. Patterns capture such relations in the form of numbers, but also in the form of sequences and they do this also in our modern society, in which much information storage and transmission is in fact entirely dependent upon such non-random patterns as barcodes.

Barcodes are a good example of patterns that carry complex information in simple form. But so are numbers, which can denote quantities, but also serve as labels and as information storage devices. Telephone numbers, for example, encode area-mapping information; while car licence plates convey a host of socially relevant information in coded form. There are many different types of number labels in the world around us, such as social security numbers, or ISBN numbers of books. The more information we process by computers, the more numbered labels are used.

The sense of order which pattern creates and assists in transmitting was first utilized in analysis by the anthropologist Claude Lévi-Strauss (1969) who drew attention to the informational content of sand-drawing from Vanuatu in the Pacific in his foundational work on the elementary structures of kinship, deploying these abstract drawings to conceive of novel ways to visualize the operational qualities of complex kinship systems, revolutionizing the study of kinship in anthropology. Patterns drawn in the sand are short-lived, their ephemerality drawing attention to the importance assigned in society to their performance.

There are thus two quite different effects that can be attributed to pattern: the first derives from pattern's iconic quality, which serves to codify or translate complex information into a new form; the second derives from the art of patterning itself, which serves to draw things together, often in new and varied ways. *Tivaivai*, in fact, behave very much like a sand-painting, yet not at all like a barcode, in emphasizing the performance of designing and executing a pattern, whose significance lies in how it is put together rather than in its representational capacity. Contemporary attitudes to pattern-making in Polynesia have resonances with traditional textile techniques that suggest that it is these attitudes that have informed the uptake of new materials and material technologies. Before returning to explicate the art of patterning exemplified by *tivaivai*, we will therefore turn briefly to a comparison of diverse perceptions of pattern in Polynesia and to the source of its efficacy.

ISBN 978-0-7141-2580-0

Barcodes, like the one for this book, carry complex information in a simple manner.

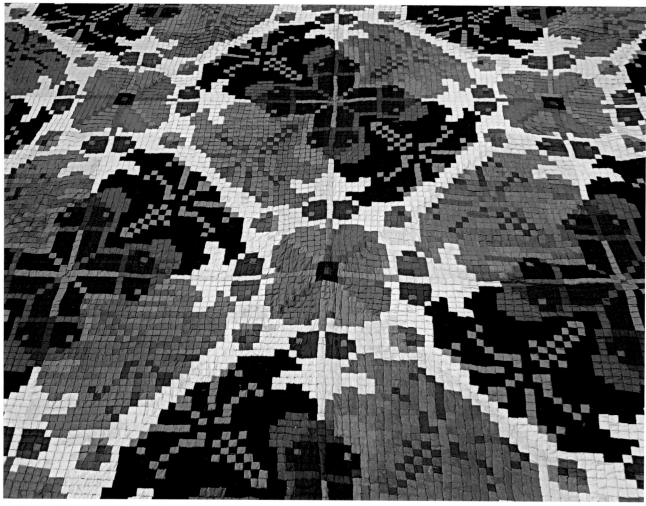

Tivaivai taorei 'Lilies', owned by Ngakura Kautai and showing the use of different fabrics in the piecing of solid colour areas.

The cultural area of Polynesia is divided into two spatial and cultural groups, notably distinguished by the continuity of barkcloth and mat production in the western part and the total replacement of barkcloth and mats as vehicles of exchange in the eastern part. The barkcloth production that continued to be practised in Western Polynesia is exemplified by a method of manufacture in which resin saps or starch are used to glue sheets of barkcloth together, creating huge extendable sheets of cloth. Eastern Polynesian barkcloth production, which came to a halt with the introduction of fabric, used a more elaborate felting method to produce continuous and yet non-extendable sheets of barkcloth. Western Polynesian barkcloth was thus produced by an additive method of composition, joining different sheets that were individually decorated with patterns using stencils, or by rubbing the bark over the raised surface pattern of a *kupesi* design board. In Western Polynesia, patterns visible on barkcloth tend to be iconically motivated, capable of mapping socially relevant information. In Eastern Polynesia, on the other hand, the felting method meant that the completed work was conceived as a whole right from the start, with patterning usually drawn freehand, or even cut into the cloth, underscoring visually distinct ways of drawing things together at a technical level, such as in the doubly layered and cut funeral cloak from the Cook Islands.

What we have here are two distinct approaches to creating a material assemblage, which, because of the pattern's importance in mapping time and history, speak eloquently about ways of doing and thinking that have remained remarkably resistant to change. In Eastern Polynesia, where *tivaivai* came to reign as the pre-eminent media for exchanges, greater importance was directed to the execution of patterning and the involvement of the material in that execution, than to the iconic potential of the pattern. Self-similar and repeated motifs of abstract nature, drawn freehand, cut or stitched by hand into the fabric, thus tend to feature prominently in the patterns of latter-day barkcloth and *tivaivai* in this part of Polynesia. The visual 'tightness' of the *tivaivai* patterns in the fabric speak eloquently about what one could call an 'art of patterning', which could dispense with the laborious work of staining and painting barkcloth when new material technologies of patterning came along.

It was possibly this preference, in Eastern Polynesia, for wholes over parts, and for calculated patterning to a given scale, that promoted the take up of fabric and sewing. Pattern in Polynesian textile arts thus directs us to distinct ways of managing quantities and quantifiable relations, which extended both to the material and the social world. *Tivaivai* show how this strategy was carried over into the techniques for handling new material, as the cloth for the entire piece is

Vaine tini shows

These days many women in the Cook Islands, especially on Rarotonga, have paid jobs that contribute to their family's income. There is often not enough time left for making *tivaivai*. Some women will sew for family members or friends and neighbours, and often get rewarded for their work with food or special favours, sometimes also with money. On Atiu most women are still members of their village women's group, *vaine tini*. During the early years of my stay on Atiu, I was keen to join, because I both wanted to learn more and also share my knowledge and ideas. So I paid my small fee and became a member of the Areora Vaine Tini Sewing Group.

The proceedings are always the same. At the year's beginning, the groups decide on their programme, which is compulsory for all members. On a fixed date, often at the end of the year, they display their works in the open, if the weather is good, or in the island's large hall when it rains. This show is always an important event. Island dignitaries are invited and seated in a nicely decorated place of honour. The women dress in their colourful village uniforms, often frilly Mother-Hubbard-style dresses made of floral print material, wearing *ei katu*, crowns of flowers, on their heads. The show is accompanied by prayers, speeches and tables full of food. Photographs are taken, and lately parts of it will even feature on the evening television broadcast. Armed with notebook and pen, *vaine tini* committee members inspect the works, each marked with its owner's name, and check that all group members have finished all the items on the programme. A woman who has not complied will not be reprimanded personally. But unfinished or missing work represents a negative mark for the entire village. Of course, the other *mamas* will not be happy to find that, despite their own hard efforts, their village lost or came second because of this, and the 'culprit' will be made quite aware of this in the weeks that follow.

Some women are not only members of their village sewing group, they also belong to a church group. Whenever a pastor and his wife finish their official term on an island, usually four years, the congregation will present them with a number of farewell gifts. A *tivaivai* will certainly be amongst them. The departing pastor's wife is honoured with the privilege of deciding what the church group's sewing programme is to be. So it can happen that one has work for both the church and

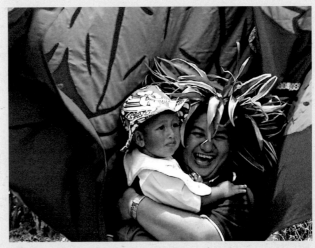

A young mother enjoying the *tivaivai* show.

village programmes; in addition to all the daily chores. The year I joined the *vaine tini* of my village, I found it so hard to complete the programme in addition to my other work that I quit the group regretfully. Since that point I have only been a spectator and admirer.

After the show is over, the *tivaivai*, pillow-cases, cushion-covers and tablecloths are wrapped in a piece of *pareu* (printed cotton cloth), and are deposited safely in a locked suitcase under the bed or a wooden glory box. Camphor balls help to keep them free of insect damage. They will live in the dark of their boxes, away from the fading sunlight, until tradition and family requirements call for their use once more.

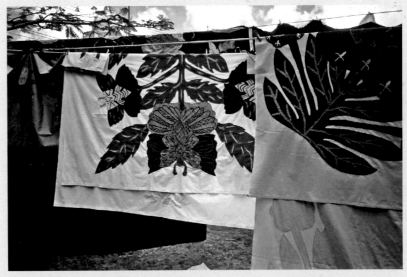

Tivaivai displayed on Atiu at the annual show in 1998.

always bought at once, demanding complex calculations to ensure just the right amount is purchased. Getting it right has practical consequences, as the colour of cloth available for purchase changes frequently. Some of the older *tivaivai taorei* (and even one or two new works) illustrate the importance of buying all the material up front, as they can show individual patches of a different hue used in a 'solid colour' area; an indication that the appropriate colour had disappeared from the shop. This illustrates the problem of availability of fabric

in remote places, a constraint that has not diminished even in modern times. In fact, one could say that the traditional material technology of Eastern Polynesian barkcloth predisposed a pattern-conscious approach to the handling of fabric which has, ironically, predisposed the technical execution to carry the full representational capacity, whereas the material and the iconic potential of the pattern remained relatively insignificant. Indicative of this emphasis on the technical potential of material is the fact that the typical

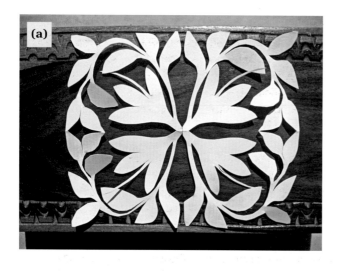

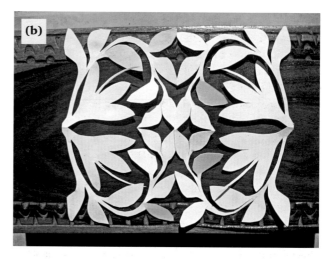

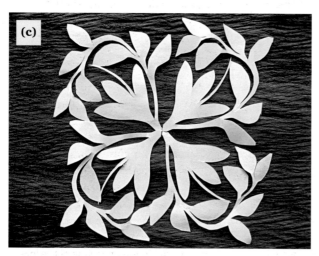

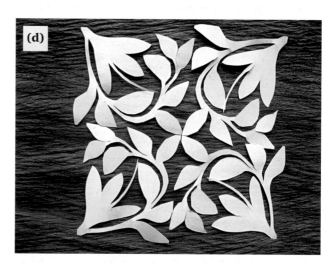

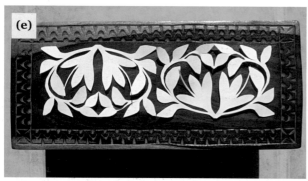

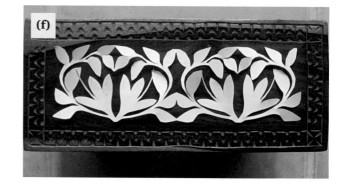

Paper cut-outs arranged in various forms of symmetry.

response to a change in fabric has been to accommodate the technique to the altered material conditions, by for example cutting larger shapes for the process of patching, rather than rebelling at the loss of quality in the fabric.

This approach to making, which consciously pre-plans the planar surface to be covered by a single pattern, cutting it in ever different ways while keeping it 'the same', has long been the hallmark of traditional Hawaiian stitching (Hammond 1986; Stewart 1986). We will now describe in some detail how this art of patterning is conceptualized and executed, in order to draw out the calculations at work in creating an overall pattern that covers the surface adequately.

In Hawaii the appliqué style used the 'snowflake' method to create a symmetrical pattern inside a square; while another method, known as piecework, was used to make geometrically repeated blocks that were added to each other to create a surface with the same overall design. The snowflake method begins with a square piece of fabric of the size to be appliquéd onto a background material. This square piece can be folded into halves, quarters, sixths, or eighths. Different, yet perfectly symmetrical, complex design patterns emerge by cutting shapes into the folded material. Folding a square into sixths creates a circular symmetrical configuration, while folding it into eighths, the most common technique, extends the patterns to the corners of the square.

Studies using computer models of the symmetric characteristics of traditional Hawaiian quilting patterns have shown that the two-dimensional plane-symmetry designs used in snowflake-type constructions in fact fall into two groups (Cao and Park 2006). These two groups illustrate the different ways of creating symmetrical patterns from a single piece of geometrically cut fabric that have become generic across all the *tivaivai*-producing cultures of Eastern Polynesia.

The first group is defined by finite or point symmetry, creating a pattern from a single geometrically cut piece of fabric, while the second group utilizes a single geometric design cut as a multiple from fabric only to be reassembled to create a fractal pattern which may be expressed in infinitely variable symmetries. Spatial transformations take place in both groups, either by rotating the pattern around a point or by reflecting it along the lines that are implicit in the symmetrical construction. While in Western-style patchwork printed fabric can be used to piece together a patterned block whose individually patterned components create a new overall composition, in *tivaivai taorei* the individual components of the *pu* (motif block) are cut from solid-colour fabric and create a visible pattern only in their totality. The arrangement and shape of the *pu* (square, triangular, circular) determine the form of symmetry, which results from the repetition of the *pu* across the surface.

Symmetries used in *tivaivai*-making result from reflection (opposite, a and b), rotation (c and d), translation (e) and glide reflection (f). It is the tension between the stasis of point symmetry in which self-similar motifs rotate around a central point, and the open-ended and expansive symmetrical construction along a line using one-, two- and multi-dimensional translation to move single motifs across a given planar surface that defines the style of *tivaivai* today across Eastern Polynesia. This style is the product not of a single invention, but of the synergic co-operation of four different types of symmetry, which bring contemporary *tivaivai* compositions into relation with one another.

Lines are animated through symmetry in the construction of a *tivaivai*, even if the lines cannot be seen, but are implicit in the construction of the pattern. There is a tension between point symmetry and the open field of fractal construction along lines. This tension impacts on the master pattern referred to as *pu* for all three *tivaivai* techniques.

Curiously, what is distinctive about *tivaivai* is neither its patterns nor its symmetrical characteristics (which may well be gleaned from pattern books that have reached the Cook Islands), but the construction of its patterns in fabric. What *tivaivai* patterns do is stitch up and connect disparate pieces of fabric to create seemingly unbreakable and visually coherent wholes. Patterning in *tivaivai* is a way of linking self-similar pieces together in ever changing ways, though the fundamentals of each *tivaivai*, their dimensions and technique of composition, remain the same.

Textile arts in the Pacific thus speak volumes about quite diverse local attitudes to patterning that in turn are hugely significant in reflecting lasting differences in ways of being and thinking. The art of *tivaivai*, and its entrance onto the world stage of art galleries and national exhibitions, showcases in its patterns one manner of making tangible ways of connecting that refuse purely cultural (but entice highly personal) modes of appropriation. It is perhaps because *tivaivai* promise to secure a visual attachment that binds a person to a pattern, that it so readily made its way into the gallery space.

Visual attachment and the question of commerce

The patterns of the *tivaivai* make visually and materially tangible the competing pull of central points and expanding fields of symmetry, which are experienced when planning and executing a design that fills the dimensions of a planar surface. Such is the analogical relation between ways of making *tivaivai* and of using them in exchange, that the commercial usage of *tivaivai* production has as yet rarely been exploited. Each and every *tivaivai* is always a very personal thing, resonating the soul of its maker as much as the time and circumstance of its making. Moving out, and returning once more, in the fashion mapped out and traced by the movements of *tivaivai*, women create the paths along which things and ideas flow back and forth between the transnational communities and the homelands. And yet sometimes *tivaivai*, as well as women, do not return.

In Hawaii and in Tahiti making patchwork for sale has been a longstanding tradition, but in the Cook Islands such commerce is less overt. While women may sell *tivaivai* to one another, or buy the sewing labour, there are few established public forums for the sale of *tivaivai*, although a number of shops now offer them on Rarotonga, though some of these are actually made in China or the Philippines. One of the few stalls on Rarotonga's daily market in Avarua, for example, that advertises in the *Cook Island Herald* that it stocks *tivaivai* for sale, mentions the textile at the very end of a long list of items that includes everything from papayas to coconuts. Approaching the stand, one will find to one's surprise that, even though no textile is visible from the outside, it harbours a rather large number of treasures. The reticence with which these fabrics are shown and offered for sale turns out to be the result of feelings that they are in fact inferior examples.

Many of the textiles on sale appear to be first attempts at sewing by younger women, showing minor faults in the execution of the patterning or roughly sewn by machine. Like the *tivaivai* that make their often repeat appearances at the many village-based competitions organized by village (or church) sewing groups (*vaine tini*), or in shows curated by the Cook Islands Tivaivai Association at the National Museum and at prestigious international textile art conventions, they have, in effect, been taken out of circulation, sometimes before they even entered the circuits of exchange in which most *tivaivai* spend their lives.

There is, however, one place that has successfully begun to sell local women's *tivaivai*, both machine- and hand-sewn. This place is the Fibre Arts Studio in Atiu, which is producing designs commercially, using motifs that are easily recognized by foreign clients, sewn in colour tones that suit the light conditions in colder climates, but in the manner

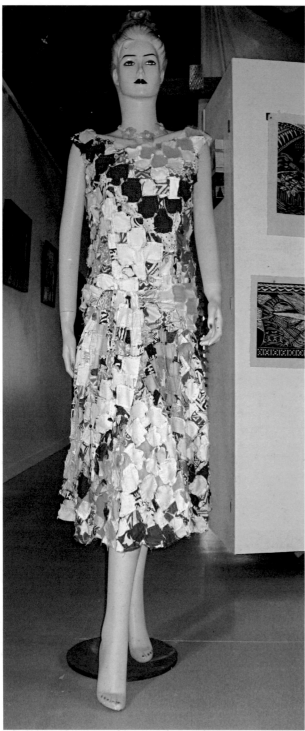

Kay George's inside-out *tivaivai taorei* dress, a wearable art work (2007).

and only an animate fabric will do when it comes to wrapping the dead and to keeping their memory alive in the living.

Yet there is a paradox, which is observed with great concern by those who have been stitching for decades: the more important *tivaivai* have become as objects of value for those who live far away from the homeland, the less time there is to make them and the less emphasis is sometimes placed on the performative aspects of making. Some have started to paint sheets of cloth, even larger than traditional *tivaivai*, as if to compensate for the reduction of time in production with a larger size of fabric.

While to some this may indicate that Cook Islands *tivaivai* are doomed, one could see this as an ingenious means of continuing the art of patterning in a manner adapted to the constraints of modern life. Ironically, however, this new way of thinking, geared towards producing objects fast and cheaply, also corrupts the efficacy of the object itself, undermining its capacity to forge connections in a social realm. Cook Islands contemporary artists have begun to counteract this development by translating the inspiration *tivaivai* provide into their own artistic language. The concept of *art* has only become part of people's consciousness on account of the South Pacific Arts Festival, which was hosted on Rarotonga in 1992. Fifteen years later a vibrant group of artists (both Cook-Islands-born and expatriate residents) has emerged, producing art that draws on both traditional Polynesian and contemporary Western ideas, materials and techniques. Their embrace of *tivaivai* as an art form came centre stage in an exhibition that the Cook Islands Tivaivai Association organized at the National Museum on Rarotonga in October 2007.

One of the leading components of the exhibition was a number of paintings taking different aspects of *tivaivai* as their subject: its history, its aesthetic quality and its social role. In her diptych *The History of Tivaivai*, Cook Islands artist Joan Rolls-Gragg provides a humorous perspective on the intertwining of the history of Christianity in Polynesia and the creation and use of *tivaivai* in Cook Islands society. Others received painterly inspiration from *tivaivai* patterns, techniques and colours. Winnie Ngaro Tau Wichman's unusual photo collages made the motifs used on *tivaivai* come to life, while P.J. Hockin painted some of them on tiny tiles; first-time exhibitor Ana Ramacake decorated her papier-mâché masks with them, and Loretta Reynolds made her own from plexiglass and *paua* shell. Kay George's *tivaivai*-style painted/printed work on plywood had the descriptive title *A Woman's Work Is Never Done*, while Krick Barraud celebrated women's skill in using the finest tissue paper as material for her delicate *tivaivai*-pattern collage *Ah-Tissue* (page 97). In *Taku Manu Nui…*, Andrea Eimke engaged in a word game, using the traditional technique of *tivaivai manu* for her stylized 'Tropic Bird' (*manu* means 'bird' in Cook Islands Maori). Colour and story-telling were prominent in the contributions of the students of Te Raranga Kite Recreational School and the private school Te Uki Ou. Miss Tiare posed next to her striking *tivaivai tataura* (embroidered appliqué) outfit (created by *tivaivai* artist Tapita Williams) which, along with her own beauty, of course, won her the title in last year's pageant, and Kay George lent her inside-out *tivaivai taorei*

of construction that is typical for Cook Islands *tivaivai*. The success of the Fibre Arts Studio shows that textile arts can be made sustainable over and above the exchange circuits in which they operate at the local level, without invalidating or undermining local concepts of value.

Machine-sewing versus hand-stitching does not seem to impart an aesthetic difference, yet with *tivaivai* it makes all the difference. The importance of the mental and physical labour of calculating the pattern, of cutting and stitching it into the fabric, animates the pattern in the minds of those who receive it and elicits the care that would be bestowed upon a living person;

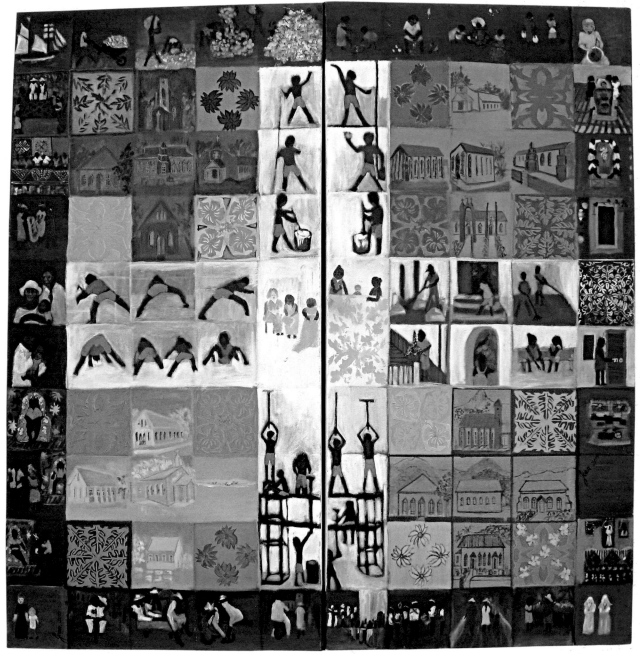

History of Tivaivai by Joan Rolls-Gragg (2007).

(mosaic patchwork) dress. In her three-piece wall quilt Jane Lamb paid homage to Cook Islands women's crafts, while Kura Tansley had her first attempts at embroidering *tivaivai* framed to perpetuate the legacy of her late embroidery teacher.

In New Zealand, artist Ani O'Neil (of Cook Islands, New Zealand and Irish descent) has created a name for herself by detaching domestic textile techniques and patterns from their homely environment and bringing into the limelight of the art world. Educated at the School of Fine Arts in Auckland in the early 1990s, she joined a group of Polynesian artists that called themselves 'The Pacific Sisters'. She became known through her performance works, the threads of which connected the lay(wo)men who created crocheted flowers. Flower networks and crocheted flowers as art objects are among her many humorous works that draw on the techniques and motifs of domesticity. One of her latest works *Lei Mai* refers to the

Pacific islands' practice of making *lei* (Hawaiian for flower wreath) as gifts to welcome visitors. The title is also a play on these words and their sound in English ('lay my'), a visual pun as she laid a large *tivaivai taorei* and cushions patterned to resemble stylized frangipani flowers that invited visitors to sit on them on the Suter Te Araroi o Whakatu Gallery's earth floor, where she was artist in residence.

Artists are leading the way in the exploration of new materials and new techniques that will continue the art of patterning in the spirit of *tivaivai*. This art, which underscores so much of the social fabric of the Cook Islands, will no doubt continue to inspire future generations to realize in their actions the lasting difference that *tivaivai* has made to their lives. One thing we know for sure: *tivaivai* will always be at the centre of the making of history in the Cook Islands.

A short history of the Atiu Fibre Arts Studio

Working with the young girls at Atiu College and seeing how much talent lay idle, knowing that many would leave the island to get a job, gave me an idea. I spoke with some local women that I knew to be experts and enthusiasts in various craft fields. They liked my suggestion of forming a company where school leavers could learn traditional and contemporary craft and artistic skills, and earn money at the same time. Seven Cook Islanders and I founded the Atiu Fibre Arts Studio, hoping that some of that year's school leavers would become our first apprentices. We found a wonderful little house, right in the middle of a romantic palm grove, that suited our requirements. I painted signboards. The *mamas* helped me plant flowers around the house. In New Zealand we found suppliers for fabrics and haberdashery, and bought sewing machines and pressing irons. Eventually the big day of the opening came. Both our members of parliament were invited, and had agreed to fly home from the main island to attend the opening ceremony. This made us feel very proud and privileged.

That day I had a lesson in cultural differences. Papa Ina, owner of our studio building and husband of one of our shareholders, and his grandson had dug an earth-oven early in the day, which was now filled with pork, chickens, taro and other good island food. At the back of the house, we had set up a large table for the meal. Each of us shareholders had jobs to do in the decoration, preparation of food or other matters. When the time approached for me to go home, get changed and to go to the airport to pick up our VIP guests everyone was still busy, but nothing was quite ready. The food was still in the oven, the *mamas* wore their stained working clothes, the table was empty! How embarrassing it would be to come back with our guests of honour to an unfinished meal! Before I lost my composure, I left and let things happen.

I drove back up from the airport as slowly as I could, feeling uncomfortable in anticipation of the chaos awaiting us. But then, to my relief, the *mamas* were there with sweet-smelling flowers in their hair and transformed into their Sunday bests, ready to greet our guests. The table was a delight of tropical food and flowers. The place looked spick and span! There had been no need to waste my energy worrying…

More useful lessons

The best two students from 'my' 1985 class, who had finished school the following year, were happy to become the first 'apprentices' at the studio. I enjoyed working with the *mamas*, who once a week showed the girls and me how to make all sorts of things using local materials such as pandanus, coconut fibre and tapa. We then set a programme. The *mamas* would work from home and bring their finished

crafts the following week. The island's total power hours had been raised from six hours daily when we arrived in 1983, to a total of twelve hours: from 5 a.m. to noon and 5–11 p.m. My two assistants and I had to plan our work carefully around these power times, using our sewing machines and pressing irons in the morning, working on hand embroideries or local crafts in the afternoon.

This wonderful work situation lasted for three months. The new year brought a new expatriate principal to the college. He had previously taught on another island in the Cooks group, and was known to be very ambitious for his students. It was this ambition that abruptly ended our idyllic set-up. He thought that all students who had left school at the end of 1986, including my assistants, should return to school and continue their education. Now I had to do all the sewing work myself, and needed to look for a new employee. I soon found what I was looking for. The principal's partner was Tahitian with Cook Islands family links. Much to my delight, she was also an extremely skilled *tivaivai* maker, and became my most important teacher. From her I learnt a special technique of machine appliqué that now forms the basis of most of our company's work.

When the principal's term finished, naturally my Tahitian teacher and colleague left with him. We had to look out for another trainee. Our local director's niece, Minna, was asked to join, and proved to be a good seamstress and interested *tivaivai* maker. We had built up stock and felt that it was time for the next step. Tourism was very limited on Atiu in those days and we needed to move, if we wanted our company to progress. We received overseas funding to mount an exhibition at an Australian university, which we were very excited about. We beat tapa and prepared bark fibres to make traditional dancing costumes. The *mamas* plaited pandanus mats, beautifully decorated with colourful patterns. Baskets were made, along with hats, purses, and shell jewellery. Our sewing machines never rested. We invested in a battery charger and converter so that we could use

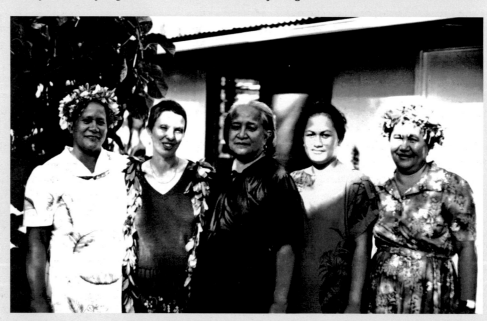

The first five founding members of Atiu Fibre Arts Studio in 1996 (left to right: Teumere Tangatapoto, Andrea Eimke, Teau Kea[†], Teremoana Paratainga, Mata Teamoke[†]).

the morning's power to charge the battery, enabling us to keep sewing in the afternoon. Many times I returned to work at the studio in the evening, so that we would have enough works to display and sell.

The day of our departure came. Our works were packed in two safe metal boxes, all the papers were duly filled in. The *mamas* decorated us with flower garlands of farewell, and away we flew to face our first overseas exhibition.

I had been totally unable to sleep with the excitement prior to our departure, and neither had Minna. When flying to Sydney from the Cook Islands, one has to stop and change planes in Auckland. The total flight time is approximately seven hours plus time waiting for the connecting flight. We were not at our best when we arrived. We briefly met with the organizers, before trying to catch some sleep. Set-up of our display was planned for the following morning, with the official opening taking place the next evening at 6 p.m.

Early the next morning we were taken to the building where everything was to happen, a large room right next to the students' dining room. The requested ironing boards and extension cords were waiting for us, we had brought all necessary equipment and we were assigned one student to help. And then we were on our own. Ten large *tivaivai* needed careful ironing, the hanging devices had to be fixed in place, the smaller items had to be displayed, and we had less than five hours before the official opening. But by 5 p.m. it all looked great. Somebody had brought in some large pot plants, and Minna and I were pleased with the result. However, we both needed a shower and enough time to transform into real women again. Outside it was pouring with rain and we had a 20-minute walk to our flat and back! When we finally returned, wet and out of breath, all the invited guests had arrived and looked at us with some surprise. The funding agency had done well, and had also arranged for a Cook Islands dance team from Sydney to perform. We were introduced to one and all, and the show began. The rest of the evening disappeared in a blur and I remember only that I narrowly escaped a nervous breakdown.

It had been my first time away from the sheltered and simple life of the islands in four years, the first time I had been exposed to the 'real world' of business. It was like being thrown into the pond without having been taught how to swim. I realized how ignorant I was of how to be a 'business woman'!

The following two weeks taught me good lessons on 'how to do my homework' and on all that was necessary for success in this context. The event turned out to be a disappointment, to a considerable extent due to my lack of expertise. One daily newspaper carried just four lines of information about the exhibition, and, though some nearby residents turned up, we made few sales, and things went very slowly. The thinner the stream of visitors, the larger my frustration. When the two weeks were finally over we had sold not even ten per cent of what we had brought with us. On our last day the liaison person for the funding agency came to see us. Even though I was thankful, of course, for their enabling us to have this exhibition, I could not help but commenting on how disappointed I was that we had sold so little. I failed to see the positive aspect of the reply, 'Be grateful that you have sold something!'

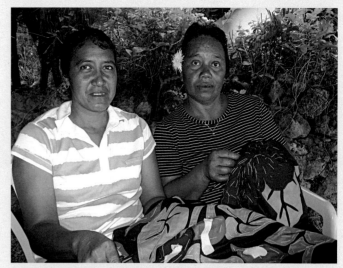

Creative colleagues since 1990, Teremoana George (left) and Tini Tivini (right) enjoy their hands-on work at the Atiu Fibre Arts Studio.

Both Minna and I were glad to return home. Back on Rarotonga, I spent time with an artist friend of mine. She sympathized when she heard our story and assured me that our 'failure' had nothing to do with the quality of our works. Slowly she helped me to rebuild my confidence. I was not going to return to Atiu with full boxes, no way! Her suggestion to mount an exhibition on Rarotonga provided the solution. And I already knew where: the Atiu Hostel.

Since becoming a self-governing nation in 1965, the Cook Islands has celebrated 'Constitution Day' every year in August. Constitution Park was built in order to stage the performances. Over the years, every island built a hostel there to accommodate their people when they come to the main island to perform in the annual celebrations. Today, all the islands have finished their flash hostels. The stage of the Constitution Park was pulled down in 1990, and the grand Cultural Centre, with its majestic auditorium, library and museum, now dwarfs the other buildings. Back then the Atiu hostel stood tall, and was one of the first and largest. It had an expansive hall overlooked by a gallery that was the ideal place to display our large *tivaivai* and craft works. It took a week to get everything organized: printing invitations and posters, driving around the island to distribute them everywhere I could. Since I had never lived on the main island, I did not know many people there, and I had no idea of who I needed to invite. The manager of the C.I. tourist authority, Te Tika Mataiapo, was extremely helpful and created a list for me that included VIPs, politicians, names of managers of hotels and companies and so forth. The fact that I had control over the organization, and the positive responses I received, raised my spirits. During the two days that I drove around the island I met many local business people. It was certainly good to take the opportunity of delivering the invitations to introduce myself and our company. Many Atiuans staying at the hostel or living on Rarotonga came to help us put up the *tivaivai*. The Big Day arrived. The display looked fabulous. Helping hands prepared and served snacks and drinks in the hostel's large kitchen. Minna and I felt happy and looked radiant. The guests arrived. Speeches were held. The refreshments were served. Our visitors were full of praise. We felt like dancing and singing. A huge weight fell from my heart. Our works were not

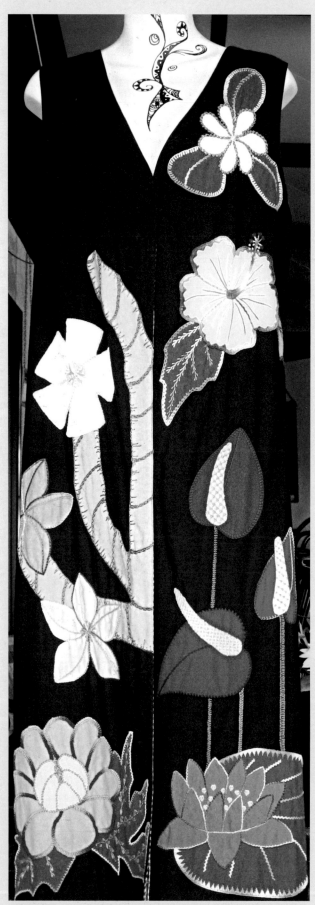

Waistcoat with *tivaivai* motifs, designed and embroidered by Andrea Eimke.

ugly, there were people out there that appreciated our talents, our efforts and our skills. This second exhibition was what Minna and I had needed in order to return to our island with our heads held high! In one week the exhibition was over, and we had sold every single item, and we returned home with empty boxes and orders in our pockets. The hard work had been worth it after all, with one exhibition a learning experience, the other a success.

More exhibitions

Encouraged by this achievement, we had an annual exhibition on Rarotonga for the following two years. Since they were held at the same time as the constitution celebrations, the Atiu Hostel was unavailable. Instead the Rarotongan Hotel, as today's Rarotongan Beach Resort and Spa was called then, let us use their conference hall. Both exhibitions were as well received and successful as the first. Business went well, but our hope of providing jobs for school leavers was not fulfilled. The younger generation still felt that the 'grass was greener on the other side of the fence'. The excitement of overseas big-city life proved more attractive than stitching *tivaivai* at home. Minna left the company to have her first child. For a while it was just 'the *mamas*' and me again, not enough hands to produce sufficient works to furnish a whole exhibition, and just enough to cater for the small tourist trade on our own island.

When Tini, a good friend and neighbour, left her job at the post office, as I knew her to be a superb embroiderer and *tivaivai* maker I asked if she would like to join me at the studio. Much to my delight, she agreed. Rather than going to the trouble of organizing our own exhibitions on Rarotonga, the following year we decided to join one organized by the C.I. Tourist authority.

An article I had written about Cook Islands *tivaivai* was published in the Australian *Textile Fibre Forum* magazine (Eimke 1989). As a result we received a request from a museum in San Francisco to help them organize a *tivaivai* exhibition. As I had planned that year to visit my family in Germany for the first time since emigrating to Atiu, I arranged the flight so that I would return home via San Francisco, and be able to talk things over with the museum's curator. I packed a couple of *tivaivai* in my suitcase as samples. She was even more enthusiastic when she saw them, and we agreed on dates and procedures. The exhibition would be held in early 1992, and I was to put a collection of *tivaivai* together and send her photographs and details of the works. This time I was determined to include other Atiu women's *tivaivai* in the exhibition, even though the idea of lending their works to be exhibited abroad as far away as the United States of America was new to them. After a lot of talking and convincing, I put together the most beautiful cross section of works possible: *tivaivai* of all kinds and colours, all heirloom pieces and none of them, of course, for sale, except the studio *tivaivai* I had shown the curator. I took slides and sent them off to San Francisco.

Mami Tepu Parua (Teau Kea), founding member and first local director in our company, had fallen ill and retired from her position. Mata Teamoke was elected our new local director. We decided that we would both fly to San Francisco for the opening of the exhibition. As daughter of the late Ngamaru Ariki, one of the three chiefs on the island, Mata had been abroad a number of times, but never as far as the USA. I had only stopped in San

Francisco for such a short time that it had not been possible to even see the museum. We had sent twelve *tivaivai*, each measuring at least 2.5 × 2.7 m, and had been informed about the small size of the museum. The curator had done wonders with props and lights, and the place shone with beauty. Two of the exhibits were Mata's: one old family heirloom piece, an exceptionally well-preserved *tivaivai taorei*, and one *tataura* she had embroidered herself.

When the television came for an interview, Mata proved a born speaker, and looked as if she had been dealing with the media all her life. The cameraman took exceptional details of the exhibits. Even though only two minutes were eventually allocated to the news item, it looked great! We received a copy of the broadcast on video to take home.

1992 was a memorable year for another reason also. The South Pacific Culture Festival is held every four years, and 1992 was the Cook Islands' turn. The decorator of the newly built Culture Centre on Rarotonga approached me with the idea of having four large (c.2 × 4 m) *tivaivai* banners made, to be hung like canopies under the high ceiling of the auditorium. It was decided that two were to be designed and hand-sewn by members of the sewing groups of all five villages; the other two would be designed and machine-sewn at the studio. Patikura Jim agreed to design the *vaine tini* (sewing group) banners, and the committees chose the best sewers from each village to work on them, their workroom being Atiu's main hall. None of us had ever worked on a *tivaivai* that large. For me it was great to see several women sewing one piece at the same time, all of them buried under the colourful masses of fabric. Unfortunately, once the banners were finished, the decorator discovered that it was impossible to access the ceiling, and the banners could not after all be hung the way she had planned. But they were incorporated into the ceremonies in other ways, and are now part of the Ministry of Culture's collection.

In the meantime my husband and I had finally managed to build our own house on Atiu. The design included a large room to be used as our workshop, so I would no longer have to drive to the studio at night if there was a need to continue working or I was suddenly overcome by inspiration. Our old home became the studio's new shop. We built a thatched half-round veranda in front of the house, and added the 'Tivaivai Café' to our shop, where people could now enjoy a cup of Atiu coffee and home-made tropical fruit cake in addition to looking at and buying our art and craft works. Sue, the wife of one of the college's expatriate teachers, was happy to escape small-island boredom by becoming our first shopkeeper. The remaining three full-time workers were glad to be able to sew without continuous interruptions from customers in our new, much more spacious, workshop.

Business was booming. Tourism was good in 1992, and the following years, on account of the publicity from the Culture Festival with its spectacular performances and its main attraction, the revival of traditional sea-voyaging skills. Our little café and studio shop on the main road proved to be a success, and we were all happy and content. The studio and its activities had been included in most of the travel guidebooks and many visitors now came to Atiu because they were curious about *tivaivai* or wanted to learn more about Atiu coffee.

A little film about tivaivai

When the late Meg Sheffield, a film producer for the BBC, visited us with the proposal of making *A Little Film About Tivaevae* I was tickled pink with excitement. What a wonderful way to introduce the colourful textile art of the Cook Islands to the world! In addition to the filming in several *tivaivai* artists' homes, I suggested we should stage a 'fake' *tivaivai* show, just like the one the members of our sewing groups hold every year in November in front of the Cook Islands Christian Church building. Cook Islanders love to be photographed or filmed, so the *mamas*' agreement was enthusiastic. There's no real celebration without food, so the BBC crew brought some ship's biscuits, cans of corned beef and soft drinks (all parts of a Cook Islander's staple diet) and the *mamas* contributed fruits and vegetables from their gardens. Soon a beautifully decorated table was set up under the tin-roofed shelter at the Boys Brigade Ground. As on real show days, washing lines criss-crossed the large yard. A fantastic selection of colourful *tivaivai* soon flapped about in the wind. Like butterflies, *mamas* in their bright village uniforms, and crowned with flowers, fluttered all over the scene. There was brilliant sunshine, a backdrop of dark blue sky to show the art at its best, and much singing, dancing and laughter. Everybody displayed the happiness the Pacific Islanders are so famous for.

It was Meg's sister, Rosie Bogle, an Auckland resident, who had planted the seed for the project by presenting her sibling with a Cook Islands *tivaivai*. Her husband, Andrew, was then the curator of the New Auckland City Art Gallery. Small wonder that the filming project brought along the idea of staging a large *tivaivai* exhibition at the gallery.

Meg and Rosie returned to the Cook Islands to show the documentary at the Atiu Nui Hall. For one evening it was transformed into a movie theatre. Meg had brought a film projector along and I provided a screen we had brought with us from Germany for slide-shows. All the members of the island's five women's groups were invited, and of course everyone came to see the film in which they had performed. Our Atiu *tivaivai* makers featured prominently in their beautifully decorated homes, but we were also thrilled to see what the other islands and the large Cook Islands community in Auckland had contributed to the film. There was much joy and laughter at the recognition of friends and relatives on other islands and, especially, in Auckland. The excellent designs and exuberant colouring of the countless *tivaivai* featured were noisily admired in the excited chatting that accompanied the screening. Meg must have felt very proud of the warm reception and never-ending applause *A Little Film About Tivaevae* received on the island of Atiu.

'Patterns of Paradise'

The film was also shown on the occasion of the colossal *tivaivai* exhibition 'Patterns of Paradise' at the New Auckland City Art Gallery in January 1995. Andrew and Rosie had gathered an enormous number of *tivaivai* not only from the islands but also from the Cook Islands community in Auckland. From businesses on our island and other sources we had raised funds to sponsor Patikura Jim, at the time our island's Women's Officer and one of the leading *ta'unga tivaivai*, to attend the opening as a representative of Atiu's *ta'unga*. I was the coordinator for the

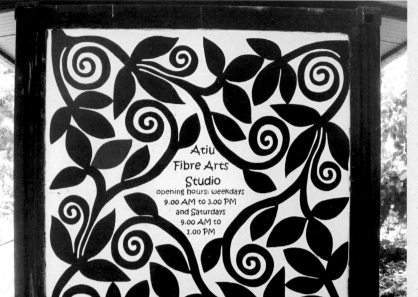

Nimble fingers quickly turned the blooms into *ei*, flower crowns for the participants and special invited guests. When Mary offered me one I thanked her, but declined. My flowers were embroidered and applied on a long waistcoat that I had made specially for the exhibition's opening. In the embroidery I used as many different stitches as I knew. The motifs were a selection of the flowers most commonly used in Cook Islands *tivaivai*. Here on Atiu it is too hot to wear such a garment. But I often use the coat as a sampler when I am looking for stitches to employ in my next embroidery project.

Tightening the belt

The 'seven fat years' had come to an end and, as the Bible tells us, the 'seven meagre years' were to follow. The world economy was on a downturn and our country, whose finances depended partly on tourism, felt the results. So did our little company. With travellers' budgets being tight, few had the money to spare for an outer

Atiu exhibits and went along with her. The women's officer from Aitutaki and the representative of the Rarotongan *ta'unga*, Mama Tokerau Munro, and her sister-in-law Kauta Dean, also managed to attend the opening. We all arrived early to lend a hand with the hanging if required. Auckland-based *tivaivai* expert Tirata Bailey and her sewing group had been commissioned to design and sew a *tivaivai* displaying the exhibition's logo. Mary Ama, an Auckland-based Cook Islander who worked as community counsellor, had been asked to dedicate her exceptional organizing skills to the project as the curator's assistant and as liaison with the *tivaivai*-making community of Auckland. Like the other Cook Islands visitors, Patikura was accommodated with family. Because I have no family in Auckland, the gallery offered me a tiny but beautiful apartment five minutes' walk from the gallery, but I had to wait a few days until it was vacant. Like a mother hen, Mary gathered me under her wings and took me home to stay with her large family. In good Cook Islands fashion, Mary's home was full of people and there were beds all over the place. I never managed to find out who was family and who was just a 'stray' passing through like myself. But her warm hospitality had room for all of us. So I tagged along with her, doing last-minute things.

There can't be a Cook Islands celebration without flowers, so the next morning Mary and I went to get some. To my amazement, Mary just went from one private house to another, and their Cook Islands inhabitants were only too happy to let her raid their gardens. We returned to the venue with arms full of the most beautiful flowers. When my financially minded '*papa'a* brain' prompted me to ask Mary how it was possible that she had been given all these flowers, she just gave me a bright smile and explained to me that it is all a give and take in a Cook Islands community. Those people either owed Mary a favour or she would now owe them one.

island stay. Sue, our first shopkeeper, had followed her husband away from the island. In Jane, a lovely Auckland-born Atiuan who had returned with her family to the home island she had never been to before, we found the ideal replacement. Young, good-looking, well trained and equipped with the warm friendliness so typical of Cook Islanders, she made a great salesperson. Her shop decorations were highly creative and kept her busy in between attending to customers. But then there were fewer and fewer customers to decorate the shop for. Turnover slowed down, commissions all but stopped. Luckily a cruise ship came to Atiu several times a year, and production continued to be worthwhile for its passengers. But the Tivaivai Café turned out to be a burden that we were no longer able to carry. We managed to find Jane other employment and transferred the studio's shop to our own home.

Luckily our new home was large. I had soon shifted things around so that our mostly unused guestroom and my bedroom could be used as the new gallery. Both rooms have a veranda in front, so the café could continue operating. It was not such a good location for a shop as the main road had been, but we soon found a solution. Our friend Rouru, who is skilled with saws, hammers and nails, built two double-sided signboards, and I applied my design skills as a sign-painter. Two neighbours agreed that we could put them up in their gardens. I had fliers printed with a map on the back showing how to find us, and deposited them at all the places tourists were likely to frequent. No longer able to rely on passing trade, our customers now had to make a special effort to find us in our rural setting. Pigs of all colours and sizes are kept in pens and tied to trees along our road. Before you arrive at our house you pass several family graveyards, and then see the studio advertised with a *tivaivai* pattern painted on the back of an old 20-foot container. But what was done out of necessity has turned out to be a benefit. The few tourists that visit us enjoy the challenge of finding our studio as an additional holiday tour. The fact that we work in the space we sell from gives visitors the opportunity to watch us at work. The shop can have much longer opening hours because we're always there to attend to our customers. And we now have much more room for displaying our work.

In the mid 1990s our little island lost half its population to migration overseas, partly because of the financial problems that had caused the government to lay off a large percentage of public servants. Rarotonga and the other islands were not much better off. Reducing our work time to half days kept our company going, but only just. But running away was not the answer.

'Artivaivai'

At the studio we had been kept busy by furnishing Are Manuiri, an alternative guesthouse in our former home. But what would happen once that work was done? As it was exactly ten years since we had had our first sales exhibition on Rarotonga, I decided it was time for another. There was now a new national museum with a high ceiling that would display *tivaivai* beautifully. The curator was only too happy to agree and we set a date for the end of the year. On the same day I met Andi Merkens for the first time. She and her family had just started living on Rarotonga, where her husband was employed as teacher for a two-year term. Andi had been trained in New Zealand as graphic artist. One of the first things she told me was how much she liked *tivaivai* and that she was keen on painting the local women and their textile art. The idea was born to have a joint exhibition with this new friend, celebrating *tivaivai* as an art form: 'Artivaivai'. From the moment I laid eyes on Andi's art I loved every single piece. She captured the atmosphere of women sewing their *tivaivai*, of the colourful house interiors, the brightness of a tropical island country. Our works complemented each other superbly. Her use of strong primary colours matched mine so perfectly that one could have thought we had discussed our works beforehand. The exhibition was well received and gave both of us a boost.

Commissions

It seemed the economy was recovering. Hotels that were up for sale changed hands, and new owners contracted companies for renovation and redecoration. One after another, our studio received contracts for *tivaivai* that would establish new and colourful looks in several hotel lobbies and offices on Rarotonga. Our sewing machines started humming again, and we could resume a full-time work schedule. The public displays helped us greatly to advertise our work.

Teaching tivaivai-*making techniques*

The editor of *Australian Textile And Fibre Forum* approached me with the request that I should teach *tivaivai*-making techniques during their annual Textile Forum in Canberra. I knew of it from the reports in the magazine, but that was all theory and these mainly displayed the results of well-established textile-art teachers' classes. I had never been to such an event, and my teaching experience was limited to two years of helping the home economics teacher at Atiu College with the fifth-form girls' embroidery project. I accepted the invitation. What a challenge!

The forum was a fascinating event. Soon I rather regretted not being there as a student, because so many renowned artists had been invited to teach their secrets. But I also enjoyed my own students, who were eager to learn the intricacies of a traditional craft most of them were unfamiliar with. I had a superb 'gofer', Lynn Inall, who was well acquainted with *tivaivai* and knew many Cook Islands *mamas* in Sydney. She would soon help to curate the wonderful *tivaivai* exhibition 'Pieces of Heaven' held at the S.H. Ervin gallery as part of the Olympic Arts Festival.

After this first step into the world of textile teaching, the next classes were less intimidating. Not only did I now accept invitations to teach *tivaivai* making abroad, I also began to offer classes at our studio workshop on Atiu. We provide individual tutoring and a relaxed atmosphere, teaching a maximum of four students at a time. Our students have come from all over the world. Often they combine the classes with a holiday.

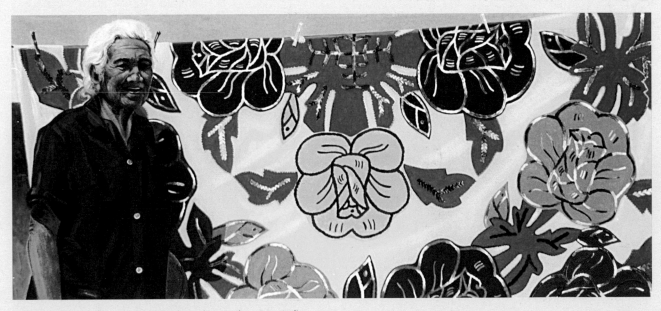

Mrs Paku and her Rose Tivaevae by Andi Merkens. Photograph courtesy Andi Merkens.

Bibliography

Akana, E., 1986, *Hawaiian Quilting: A Fine Art*. Honolulu: Hawaiian Childrens Society.

Appadurai, A., 1986, 'Commodities and the politics of value'. In A. Appadurai (ed.), *The Social Life of Things: Commodities in Cultural Perspective*: 3–63. Cambridge: Cambridge University Press.

Ascher, M., 2002, *Mathematics Elsewhere*. Princeton: Princeton University Press.

Babadzan, A., 1993, *Les Dépouilles des Dieux: Essai sur la Religion Tahitienne à l'Epoque de la Découverte*. Paris: Editions de la Maison des Sciences.

——— 2003, 'The gods stripped bare'. In C. Colchester (ed.), *Clothing the Pacific*: 25–51. Oxford: Berg.

Beaglehole, E., 1948, 'Social and political changes in the Cook Islands', *Pacific Affairs* 21(4): 384–98.

——— 1957, *Social Change in the South Pacific: Rarotonga and Aitutaki*. London: George Allen & Unwin Ltd.

Belting, H., 1995, 'Global art and minorities: a new geography of art history'. In H. Belting (ed.), *Art History after Modernism*: 62–74. Chicago: Chicago University Press.

Bennardo, G., 2002, *Representing Space in Oceania: Culture in Language and Mind*. Canberra: Australian National University.

Berlin, B. and Kay, P., 1969, *Basic Color Terms: Their Universality and Evolution*. Berkeley and Los Angeles: University of California.

Bolton, L., 2003, 'Gender, status and introduced clothing in Vanuatu'. In C. Colchester (ed.), *Clothing the Pacific*: 119–39. London: Berg Publishers.

——— 2007, '"Island dress that belongs to us all": mission dresses and the innovation of tradition in Vanuatu'. In E. Ewart and M. O'Hanlon (eds), *Body Arts and Modernity*: 165–82. Wantage: Sean Kingston Publishing.

Borofsky, R., 1987, *Making History: Pukapukan and Anthropological Constructions of Knowledge*. Cambridge: Cambridge University Press.

Buck, P.H. (Te Rangi Hiroa), 1927, *Material Culture of the Cook Islands (Aitutaki)*: New Plymouth: Ams Pr Incorporated.

——— 1944, *Arts and Crafts of the Cook Islands*, B.P. Bishop Museum Bulletin 179.

——— 1945, *Introduction to Polynesian Anthropology*, B.P. Bishop Museum Bulletin 187: 92–6.

Burdick, N., 1988, *Legacy: The Story of Talula Gilbert Bottoms and her Quilts*. Nashville: Rutledge Hill.

Buse, J. and Taringa, R., 1996, *Cook Islands Maori Dictionary*, Research School of Pacific and Asian Studies, The Australian National University.

Cao, T. and Park, J.H., 2006, 'Symmetric characteristics of traditional Hawaiian patterns: a computer model'. In R. Sahangi and J. Sharp (eds), *Brides London, Mathematics, Music, Art, Architecture, Culture*: 89–98. Conference Proceedings, printed by Print Solutions Partnership.

Colchester, C. (ed.), 2003, *Clothing the Pacific*. Oxford: Berg.

Crocombe, R.G., 1964., *Land Tenure in the Cook Islands*. Melbourne: Oxford University Press.

——— (ed.), 1987, *Land Tenure in the Atolls: Cook Islands, Kiribati, Marshall Islands, Tokelau, Tuvalu*. Suva, Fiji: Institute of Pacific Studies.

Crocombe, R., and Tua'inekore Crocombe, M. (eds), 2003, *Akono'anga Maori Cook Islands Culture*. Suva, Fiji: Institute of Pacific Studies.

Daston, L., 2004, 'Glass flowers', In L. Daston (ed.) *Things that Talk: Object Lessons from Art and Science*: 223–57. New York: Zone Books.

Davis, T., 1992, *Island Boy – An Autobiography*. Suva, Fiji: Institute of Pacific Studies.

Eimke, A., 1989, 'Polynesian Tifaifai', *Australian Textile And Fibre Forum* 8(2), no. 25: 8–9.

Elsley, J., 1996, *Quilts as Text(iles): The Semiotics of Quilting*. Los Angeles: Berkeley: University of California Press.

Evans, M., 2001, *Persistence of the Gift: Tongan Tradition in Transnational Context*. Waterloo, Ont.: Wilfred Laurier University Press.

Eves, R., 1996, 'Remembrance of things passed: memory, body and the politics of feasting in New Ireland, Papua New Guinea', *Oceania* 66(4): 257–327.

Fodor, J., 1998, *Concepts: Where Cognitive Science Went Wrong*. Oxford: Oxford University Press.

Freedberg, D., 1989, *The Power of Images: The Study in the History and Theory of Response*. Chicago: Chicago University Press.

Gell, A., 1985, 'How to read a map: remarks on the practical logic of navigation', *Man* (NS) 20(2): 271–86.

——— 1992, *The Anthropology of Time*. Oxford: Oxford University Press.

——— 1996, 'Vogel's net: traps as artworks and artworks as traps', *Journal of Material Culture* 1(1): 15–39.

——— 1998, *Art and Agency: An Anthropological Theory*. Oxford: Oxford University Press.

Gill, W., 1856, *Gems from the Cook Islands*, vol. 2. London.

Gilson, R., 1980, *The Cook Islands 1820–1950* (ed. R. Crocombe). Wellington: Victoria University of Wellington.

Gosset, R.W., 1940, 'Notes on the Discovery of Rarotonga', *Australian Geographer* 3: 41–51

Granet, M., 1968, *Le Pensée chinoise*. Paris: Editions Albin Michel.

Hacking, I., 1990, *The Taming of Chance*. Cambridge: Cambridge University Press.

Hammond, J., 1986, 'Polynesian women and tifaifai fabrications of identity', *The Journal of American Folklore* 99(393): 259–79.

Hanson, F.A., 1983, 'When the map is a territory: art in Maori culture'. In D.K. Washburn (ed.), *Structure and Cognition in Art*. Cambridge: Cambridge University Press.

Heider, K. and Rosch, E., 1972, 'Universals in color naming and memory', *Journal of Experimental Psychology* 93(1): 10–20.

Henare, A., Holbraad, M and Wastell, S. (eds), 2007, *Thinking Through Things*. Cambridge: Cambridge University Press.

Henry, T., 1928, *Ancient Tahiti*, Bernice Bishop Museum Bulletin 48: 158–60.

Horton, L., 2005, *Mary Black's Family Quilts: Memory and Meaning in Everyday Life*. Tennessee: University of South Carolina Press.

Howes, D., 1987, 'Olfaction and transition: an essay on the ritual uses of smell', *Canadian Review of Sociology and Anthropology* 24: 398–410.

——— 2003, *Sensual Relations*. Ann Arbor: University of Michigan Press.

Hutchins, E., 1983, *Mental Models*. New York: Lawrence Erlbaum Associates.

Ingold, T., 2000, 'Making culture and weaving the world'. In P.M. Graves-Brown (ed.), *Matter, Materiality, and Material Culture*: 50–72. London: Routledge.

——— 2007. *Lines: A Brief History*. Oxford: Berg.

Johnson, M., 1987, *The Body in the Mind: The Bodily Basis Of Meaning, Imagination and Reason*. Chicago: University of Chicago Press.

Jones, S., 1973, *Hawaiian Quilts*. Honolulu: Hawaii University Press.

Kamil, N., 2005, *The Fortress of the Soul: Violence, Metaphysics and Material Life in the Huguenot's New World, 1517–1751*. Baltimore: Johns Hopkins University Press.

Kaeppler, A.L., 1978, 'Melody, drone and decoration: underlying structures and surface manifestations in Tongan art and society'. In M. Greenhalgh and V. Megaw (eds), *Art in Society: Studies in Style, Culture and Aesthetics*: 137–78. London: Gerald Duckworth.

Kautai, N., Malcolm, T.K., Mokoroa, P., Tanga, T., Tanga, T., Tangatapoto, V., Tatuava, T. and Touna, T., 1984, *Atiu – An island community*. Suva, Fiji: Institute of Pacific Studies.

Keane, W., 2005, 'The hazards of new clothes: what signs make possible'. In S. Küchler and G. Were (eds), *The Art of Clothing*: 1–19. London: University College London and Cavendish Press.

Kingston, S., 2003, 'Form, attention and a southern New Ireland life cycle', *Journal of the Royal Anthropological Institute* 9(4): 681–708.

Klein, J., 1997, 'Matisse after Tahiti – the domestication of exotic memory', *Zeitschrift für Kunstgeschichte* Band 60, Heft 1: 44–89.

Kloosterman, A.M.J., 1976, *Discoverers Of The Cook Islands and the Names they Gave* (second revised edition). Rarotonga, Cook Islands: Cook Islands Library and Museum Bulletin 1.

Kooijman, S., 1972, *Tapa in Polynesia*, Bernice P. Bishop Museum Bulletin 234. Honolulu, Hawaii: Bishop Museum Press.

Küchler, S., 2002, 'Imaging the body politic: the knot in Pacific imagination', *L'Homme* 165: 205–33.

——— 2003, 'The poncho and the quilt: material Christianity in the Cook Islands'. In C. Colchester (ed.), *Clothing the Pacific*: 97–119. Oxford: Berg.

——— 2004. 'On art and mathematics' (in A. Barrowclout (ed.), 'Art and Archaeology: Unmasking Material Culture'), *Archaeological Review from Cambridge* 19(1): 28–46.

——— 2005, 'The modality of time-maps: quilting from another point of view', *RES: Journal of Anthropology and Aesthetics* 47: 179–90.

Küchler, S. and Were, G., 2005, *Pacific Pattern* (with photographer G. Jowitt). London: Thames & Hudson.

Landes, D.S., 1983, *Revolution in Time*. Massachusetts: Belknap Press of Harvard University.

Lehman, F.K. and Herdrich, D.J., 2002, 'On the relevance of point field for spatiality in Oceania'. In Giovanni Bennardo (ed.), *Representing Space in Oceania* (Culture in Language and Mind No 523): 179–97. Canberra: Pacific Linguistics.

Lee, I., 1920, *Captain Bligh's Second Voyage to the South Sea*. London: Longmans, Green & Co.

Lévi-Strauss, C., 1969, *The Elementary Structures of Kinship* (trans. J.H. Belle, J.R. von Sturmer and R. Needham). Boston: Beacon Press.

Lippard, L.R., 1997, *The Lure of the Local: Senses of Place in a Multicentered Society*. New York: New Press.

Maretu, 1983, *Cannibals and Converts - Radical change in the Cook Islands* (trans., ed. and annot. M.T. Crocombe). Institute of Pacific Studies University of the South Pacific in association with the Ministry of Education, Rarotonga.

Marston, G. and Cunningham, J., 1990, *Mary Schafer and her Quilts*. East Lansing: Michigan State University Museum.

Mason, J.T., 2003, 'Cultural Influences from and on inter-cultural marriage'. In R. Crocombe and M. Tua'inekore Crocombe (eds), *Akono'anga Maori: Cook Islands Culture*: 247–61. Suva, Fiji: Institute of Pacific Studies.

Mauss, M., 1967 [1923], *The Gift: Forms and Functions of Exchange in Archaic Societies* (trans. Ian Cunnison). New York: The Norton Library.

Mead, M., 1978, *Brombeerblüten im Winter*. Reinbek: Rowohlt Taschenbuch Verlag.

Metge, J., 1976, *The Maoris of New Zealand*. London, Boston: Routledge & Kegan Paul.

Moss, F.J., 1894, 'Maori polity of the islands of Rarotonga', *Journal of the Polynesian Society* 3: 20–6.

Mukerji, C., 1983, *From Graven Images: Patterns of Modern Materialism*. New York: Columbia Press.

Neich, R. and Pendergrast, M., 1997, *Traditional Tapa Textiles of the Pacific*. London: Thames & Hudson.

O'Bagy Davis, C., 1990, *Pioneer Quiltmaker: The Story of Dorinda Moody Slade, 1805–1895*. Tucson: Sanpete.

Oliver, D. 2002, *On Becoming 'Old' in Early Tahiti and Early Hawaii: A Comparison*. Papeete, Tahiti: Editions Haere Po.

Pakoti, J., 1895, 'First inhabitants of Aitutaki', *Journal of the Polynesian Society* 4: 59–70.

Parker, R., 1984, *The Subversive Stitch: Embroidery and the Making of the Feminine*. London: Women's Press.

Phillips, R., 1999, *Trading Identities: The Souvenir in Native North American Art from the North East 1700–1900*. Seattle: University of Washington Press.

Pogglioli, V., 1988, *Patterns from Paradise: The Art of Tahitian Quilting*. Pittstown, NJ: Main Street Press.

Rampley, M., 2005, 'Art history and cultural difference: Alfred Gell's anthropology of art', *Art History* 28(4): 524–51.

Rochette, M., 2003, 'On the meanings of Burekalou, a.k.a. "Model Spirit House" from Fiji', *RES: Anthropology and Aesthetics* 44: 70–98.

Rongokea, L., 2001 [1992], *The Art of Tivaevae: Traditional Cook Islands Quilting*. Honolulu: University of Hawaii Press.

Rubinstein, D. (ed.), 1992, *Pacific History: Papers from the 8th Pacific History Association Conference*. Guam: University of Guam Press.

Savage, S., 1980 [1962], *A Dictionary of the Maori Language of Rarotonga (1890s – 1962)*. Suva, Fiji: Institute of Pacific Studies, University of the South Pacific in association with the Cook Islands Ministry of Education.

Schiebinger, L. and Swan, C. (eds), 2005, *Colonial Botany, Science, Commerce and Politics in Early Modern Europe*. Philadelphia: University of Pennsylvania Press.

Schoeser, M., 2003, *World Textiles: A Concise History*. London: Thames & Hudson.

Semper, G., 2004, *Style in the Technical and Tectonic Arts; or, Practical Aesthetics* (trans. H.F. Mallgrave). Santa Monica: The Getty Publications.

Sennett, R., 1974, *The Fall of Public Man*. New York: W.W. Norton.

Shapiro, H.L. and Buck, P.H., 1936, *Physical Characters of the Cook Islanders* (Memoirs vol. 22, no 1). Honolulu, Hawaii: P.B. Bishop Museum.

Shaw, R., 1996, *Hawaiian Quilt: Masterpieces*. New York : Hugh Lauter Levin Associates.

Sheffield, M., 1995, *A Little Film About Tivaevae*. BBC.

Siikala, J., 1991, *Akatokamava: Myth, History and Society in the Southern Cook Islands*. Auckland: The Polynesian Society in Association with the Finnish Anthropological Society, Helsinki.

Silverman, E., 1998, *Times Enmeshed: Gender, Space and History Among the Duna of Papua New Guinea*. Stanford, CA: Stanford University Press.

Smith, P. and Findlen, P. (eds), 2002, *Merchants and Marvels: Commerce, Science and Art in Early Modern Europe*. New York: Routledge.

Snyder , G., 1986 [1963], *No Time On My Hands*. Lincoln: University of Nebraska Press.

Spurling, H., 2005, *The Unknown Matisse*. New York: Knopf.

Stafford, B.M., 2007, *Echo Objects: The Cognitive Work of Images*. Chicago: Chicago University Press.

Stewart, S., 1986, *Snowflakes in the Sun*. Pennsylvania: Wallace-Homestead Books.

Strathern, M., 1988, *The Gender of the Gift: Problems with Women and Problems with Society in Melanesia*. Berkeley: University of California Press.

Sunderland, J.P. and Buzacott, A., 1866, *Mission Life in the Islands of the Pacific*. London: John Snow & Co.

Tanga, T., 1984, 'Enua – The island'. In N. Kautai *et al.* (eds), *Atiu – An Island Community*. Suva, Fiji: Institute of Pacific Studies of the University of the South Pacific.

Tcherkézoff, S., 2003, 'On cloth, gifts and nudity: Regarding some European misunderstandings during early encounters in Polynesia'. In C. Colchester (ed.), *Clothing the Pacific*: 51–79. Oxford: Berg.

Te Rangi Hiroa, 1939, *Anthropology and Religion*. New Haven: Yale University Press.

Thomas, N., 1995, *Oceanic Art*. London: Thames & Hudson.

——— 1996a , 'Te Moemoea no Iotefa/The Dream of Joseph', *The Contemporary Pacific* 8(2): 291–317.

——— 1996b, 'Cold Fusion', *American Anthropologist* 38(1): 9–16.

——— 1999, 'The case of the misplaced ponchos: speculations concerning the history of cloth in Polynesia', *Journal of Material Culture* 4(1): 5–21.

——— 2006, 'Our history is written in our mats: reflections on contemporary art, globalisation and history'. In L. Seear and S. Raffel (eds), *The 5th Asia-Pacific Triennial of Contemporary Art*: 24–32. Brisbane: Queensland Art Galley Publishing.

Thoreau, H.D., 1971 [1854], *Walden*. New Jersey: Princeton University Press.

Tobin, J.L. and Dobard, R.G., 1999, *Hidden in Plain View: The Secret Story of Quilts and the Underground Railroad*. New York: Anchor Books.

Valeri, V., 1985, *Kingship and Sacrifice in Ancient Hawaii*. Berkeley: University of California Press.

Wagner, R., 1992, 'The fractal person'. In M. Godelier and M. Strathern (eds), *Big Men and Great Men: Personifications of Power in Melanesia*. Cambridge: Cambridge University Press.

Washburn, D. and Crowe, D., 1988, *Symmetry of Culture: Theory and Practice of Plane Pattern Analysis.* Seattle: University of Washington Press.

——— 2004, *Symmetry Comes of Age: The Role of Pattern in Culture.* Seattle: University of Washington Press.

Wassmann, J., 1994, 'The Yupno as post-Newtonian scientists: the question of what is "natural" in spatial conception', *Journal of the Royal Anthropological Institute* 29(3): 645–66.

Weiner, A., 1992, *Inalienable Possessions: The Paradox of Keeping-while-Giving.* Berkeley: University of California Press.

——— 1994, 'Cultural difference and the density of objects', *American Ethnologist* 21(2): 391–403.

Welsch, R.L., Venbrux, E. and Rosi, P.S., 2006, 'Exploring world art: an introduction'. In *Exploring World Art*: 1–37. Long Grove, Illinois: Waveland Press.

Were, G., 2005, 'Pattern, efficacy and enterprise: on the fabrication of connections in Melanesia'. In S. Küchler and D. Miller (eds), *Clothing as Material Culture*: 159–175. Oxford: Berg.

——— in press. *Lines that Connect.* Honolulu: Hawaii University Press.

Williams, J., 1938, *A Narrative of Missionary Experience.* London: John Snow & Co.

Wragge, C.L., 1906, 'The romance of the South Seas'. London: Chatto & Windus.

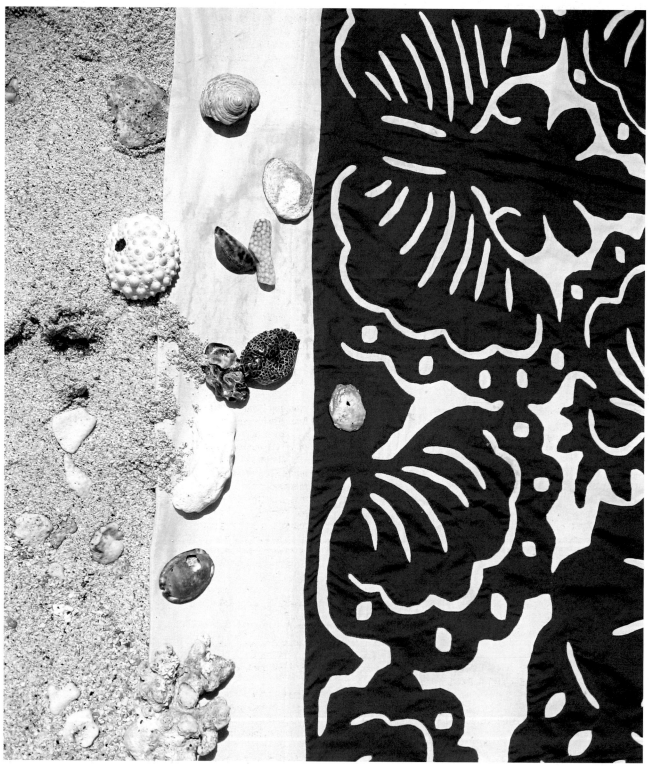

'Conch and Clam Shells', a *tivaivai manu* owned and sewn by Tangata Edwards (Atiu) and designed by Patikura Jim.